11+

Non-Verbal Reasoning

Practice Papers

2

Hachette UK's policy is to use papers that are natural, renewable and recyclable products and made from wood grown in well-managed forests and other controlled sources. The logging and manufacturing processes are expected to conform to the environmental regulations of the country of origin.

Orders: please contact Hachette UK Distribution, Hely Hutchinson Centre, Milton Road, Didcot, Oxfordshire, OX11 7HH. Telephone: (44) 01235 400555. Email: primary@hachette.co.uk. Lines are open from 9 a.m. to 5 p.m., Monday to Friday.

Parents, Tutors please call: 020 3122 6405 (Monday to Friday, 9:30 a.m. to 4.30 p.m.). Email: parentenquiries@galorepark.co.uk

Visit our website at www.galorepark.co.uk for details of revision guides for Common Entrance, examination papers and Galore Park publications.

ISBN: 978 1 471869 07 5

Typeset in India

Illustrations by Peter Francis

Printed in the UK

A catalogue record for this title is available from the British Library.

Name: _____

11+

Non-Verbal Reasoning

Practice Papers

2

Peter Francis & Sarah Collins

GALORE PARK

AN HACHETTE UK COMPANY

Contents and progress record

How to use this book   6

Practice Papers 1 contains papers 1–8 and should be attempted first

Ideal for...	Paper	Page	Length	Timing
General training for all 11+ and pre-tests, good for familiarisation of test conditions.	1	9	20	13:30
	2	15	20	13:30
	3	19	20	10:00
	4	24	20	13:30
Short-style tests designed to increase speed. Good practice for pre-test and CEM 11+.	5	28	20	10:00
	6	33	20	13:30
	7	37	30	11:00
	8	44	10	05:00

Practice Papers 2 contains papers 9–13 and should be attempted after *Practice Papers 1*

Ideal for...	Paper	Page	Length (no. Qs)	Timing (mins)
Long-style tests, good practice for GL.	9	9	60	30:00
	10	18	60	22:00
Varied-length tests with challenging content. Good practice for pre-test, CEM, GL or any independent school exam.	11	27	30	11:00
	12	35	10	05:00
	13	37	60	22:00

Answers   49

Speed	Score		Time		Notes
Slow		/ 20	:		
Slow		/ 20	:		
Average		/ 20	:		
Average		/ 20	:		
Average		/ 20	:		
Average		/ 20	:		
Fast		/ 30	:		
Fast		/ 10	:		

Speed	Score		Time		Notes
Average		/ 60	:		
Fast		/ 60	:		
Fast		/ 30	:		
Fast		/ 10	:		
Fast		/ 60	:		

How to use this book

Introduction

These *Practice Papers* have been written to provide final preparation for your 11+ Non-Verbal Reasoning test.

Practice Papers 1 includes eight model papers with a total of 160 questions. There are ...

- four training tests, which include some simpler questions and slower timings designed to develop your confidence
- four tests in the style of Pre-Tests and short-format CEM (Centre for Evaluation and Monitoring)/bespoke tests in terms of difficulty, speed and question variation.

The tests cover the standard Non-Verbal Reasoning questions and spatial reasoning questions (where you are asked to imagine a picture from a different angle or perspective), including 3D shapes.

Practice Papers 2 includes five model papers with a total of 220 questions. These papers include ...

- longer-format tests in the GL (Granada Learning)/bespoke style
- a spatial reasoning test with challenging content, speed and question variation to support all 11+ tests.

Practice Papers 2 will help you ...

- become familiar with the way long-format 11+ tests are presented
- work on more short-format papers
- build your confidence in answering the variety of questions set
- work with the most challenging questions set
- tackle questions presented in different ways
- build up your speed in answering questions to the timing expected in the most demanding 11+ tests.

Pre-Test and the 11+ entrance exams

The Galore Park 11+ series is designed for Pre-Tests and 11+ entrance exams for admission into independent schools. These exams are often the same as those set by local grammar schools too. Non-Verbal Reasoning tests now appear in different formats and lengths and it is likely that if you are applying for more than one school, you will encounter more than one type of test. These include:

- Pre-Tests delivered on-screen
- 11+ entrance exams in different formats from GL (Granada Learning) and CEM (Centre for Evaluation and Monitoring)
- 11+ entrance exams created specifically for particular independent schools.

Tests are designed to vary from year to year. This means it is very difficult to predict the questions and structure that will come up, making the tests harder to revise for.

To give you the best chance of success in these assessments, Galore Park has worked with 11+ tutors, independent school teachers, test writers and specialist authors to create these *Practice Papers*. These books cover the styles of questions and the areas of Non-Verbal Reasoning that typically occur in this wide range of tests.

*Because 11+ tests now aim to include variations in the content and presentation of questions, making them increasingly difficult to revise for, **both** books should be completed as essential preparation for all Pre-Test and 11+ Non-Verbal Reasoning tests.*

For parents

These *Practice Papers* have been written to help both you and your child prepare for Pre-Test and 11+ entrance exams.

For your child to get the maximum benefit from these tests, they should complete them in conditions as close as possible to those they will face in the exams, as described in the 'Working through the book' section below.

Timings get shorter as the book progresses to build up speed and confidence.

Some of these timings are very demanding and reviewing the tests after completing the books (even though your child will have some familiarity with the questions) can be helpful, to demonstrate how their speed has improved through practice.

For teachers and tutors

This book has been written for teachers and tutors working with children preparing for both Pre-Test and 11+ entrance exams. The variations in length, format and range of questions are intended to prepare children for the increasingly unpredictable tests encountered, with a range of difficulty developed to prepare them for the most challenging paper and on-screen adaptable tests.

Working through the book

The **Contents and progress record** helps you to understand the purpose of each test and track your progress. Always read the notes in this record before beginning a test as this will give you an idea of how challenging the test will be!

You may find some of the questions hard, but don't worry. These tests are designed to build up your skills and speed. Agree with your parents on a good time to take the test and set a timer going. Prepare for each test as if you are actually going to sit your 11+ (see 'Test day tips' on page 8).

- Complete the test with a timer, in a quiet room. Note down how long it takes you and write your answers in pencil. Even though timings are given, you should complete ALL the questions.
- Mark the test using the answers at the back of the book.
- Go through the test again with a friend or parent and talk about the difficult questions.
- Have another go at the questions you found difficult and read the answers carefully to find out what to look for next time.

The **Answers** are designed to be cut out so that you can mark your papers easily. Do not look at the answers until you have attempted a whole paper. Each answer has a full explanation so you can understand why you might have answered incorrectly.

When you have finished a test, turn back to the 'Contents and progress record' and fill in the boxes:

- Write your total number of marks in the 'Score' box
- Note the time you took to complete ALL the questions in the 'Time' box.

After completing both books you may want to go back to the earlier papers and have another go to see how much you have improved!

Continue your learning journey

When you've completed these *Practice Papers*, you can carry on your learning right up until exam day with the following resources.

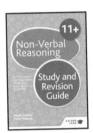 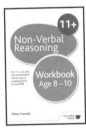 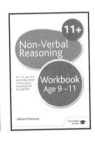 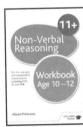   

The *Study and Revision Guide* is an essential tool in your 11+ preparation. The book explains how to tackle the different types of questions you may encounter in the Non-Verbal Reasoning 11+ entrance exams and ways to use your knowledge of Maths to break down the questions easily. Using the *Study and Revision Guide* as part of your 11+ preparation can significantly improve your speed in answering questions.

Practice Papers 1 contains four training and four short-format papers and answers to increase your familiarity with taking actual tests and to improve your accuracy, speed and ability to deal with a wide range of questions under pressure.

CEM 11+ Non-Verbal Reasoning & Spatial Reasoning Practice Papers contains three practice papers designed for preparation for the CEM-style tests. Each paper is split into short tests in non-verbal reasoning and spatial reasoning. The tests vary in length and format and are excellent for short bursts of timed practice.

GL 11+ Non-Verbal Reasoning Practice Papers contains three practice papers designed for preparation for the GL-style tests.

The *Workbooks* will develop your skills, with over 160 questions to practise in each book. To prepare you for the exam, these books include even more question variations that you may encounter – the more you do, the better equipped for the exams you'll be.

- *Age 8–10*: Increase your familiarity with variations in the question types.
- *Age 9–11*: Experiment with further techniques to improve your accuracy.
- *Age 10–12*: Develop fast response times through consistent practice.

Paper 9

Look at the first three pictures and decide what they have in common. Then select the option from the five on the right that belongs to the same set. Circle the letter beneath the correct answer. For example:

a   ⓑ   c   d   e

1

a   b   c   d   e

2

a   b   c   d   e

3

a   b   c   d   e

4

a   b   c   d   e

5

a   b   c   d   e

6

a   b   c   d   e

7

a   b   c   d   e

Turn over to the next page.

8

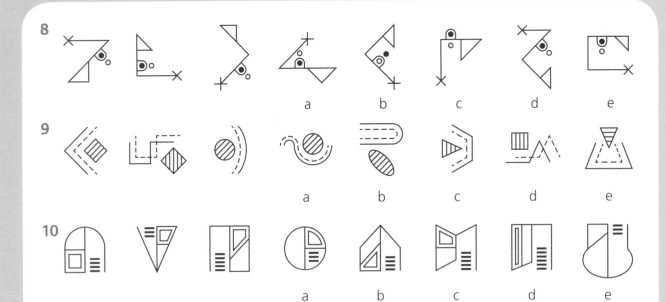

a   b   c   d   e

9

a   b   c   d   e

10

a   b   c   d   e

Look at the two pictures on the left connected by an arrow. Decide how the first picture has been changed to create the second. Now apply the same rule to the third picture and circle the letter beneath the correct answer. For example:

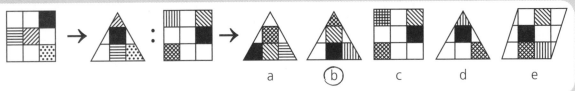

a   (b)   c   d   e

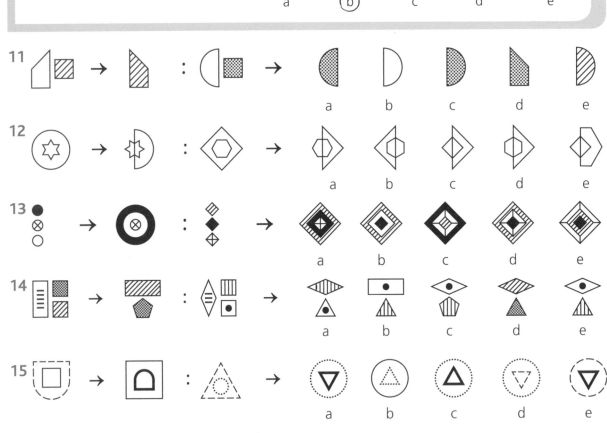

11

a   b   c   d   e

12

a   b   c   d   e

13

a   b   c   d   e

14

a   b   c   d   e

15

a   b   c   d   e

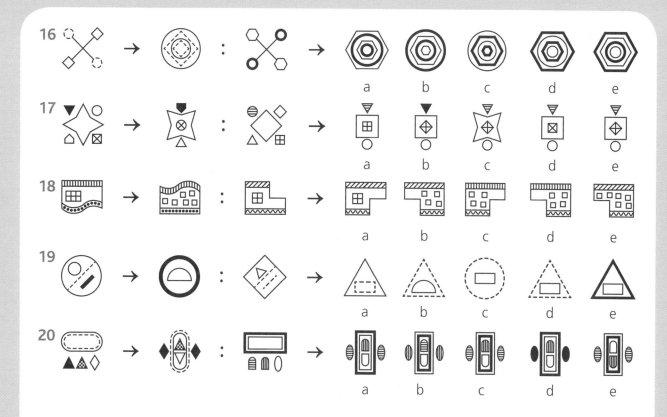

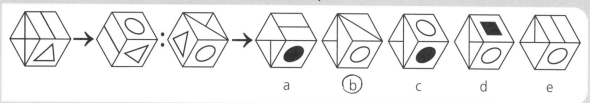

The first cube has been rotated to a new position, shown after the arrow. One of the answer options shows the third cube, rotated in the same way as the first cube. Circle the letter beneath the correct answer. For example:

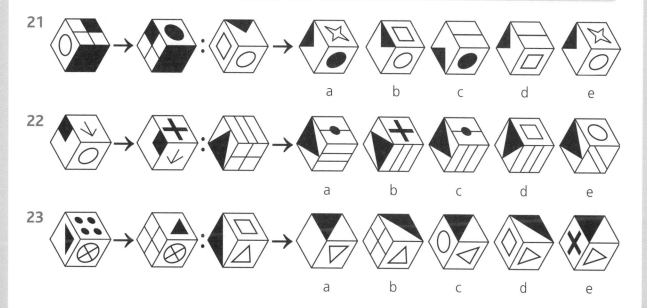

Turn over to the next page.

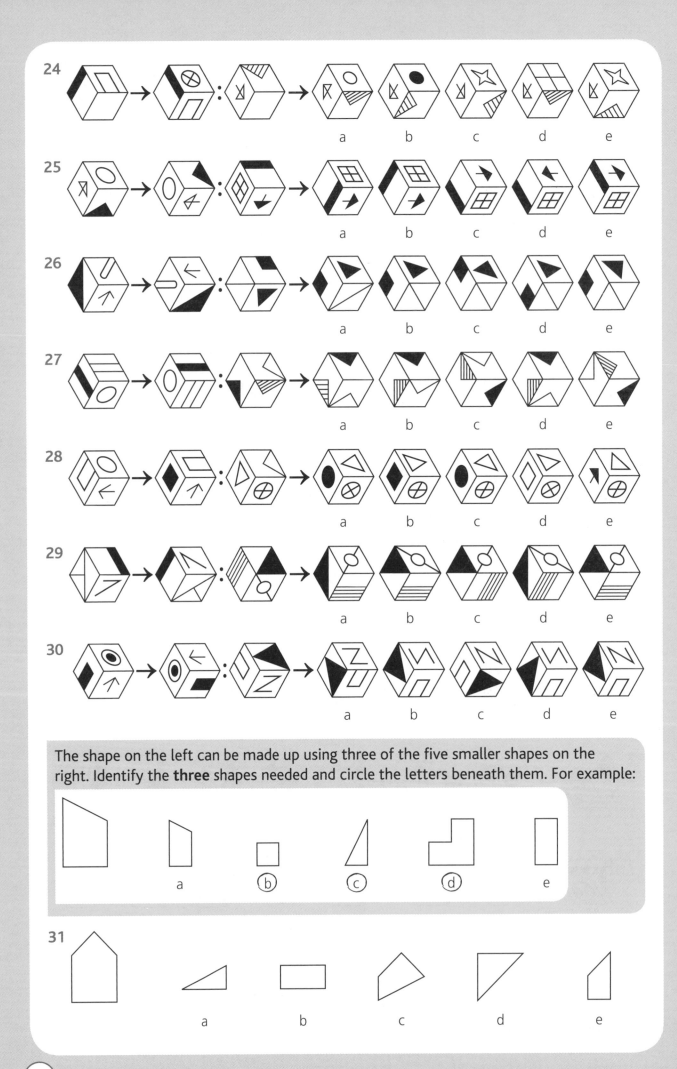

24

25

26

27

28

29

30

The shape on the left can be made up using three of the five smaller shapes on the right. Identify the **three** shapes needed and circle the letters beneath them. For example:

a   (b)   (c)   (d)   e

31

a   b   c   d   e

32 a b c d e

33 a b c d e

34 a b c d e

35 a b c d e

36 a b c d e

37 a b c d e

38 a b c d e

39 a b c d e

40 a b c d e

Turn over to the next page.

Each letter represents an individual feature in the picture next to it. Work out which feature is represented by each letter. Apply the code to the picture in the box and circle the letter beneath the correct answer code. For example:

 SUW

 TVX

 TUY

 SVZ

TVZ	SUY	SVX	SUW	TUZ
a	b	c	d	(e)

41 RX

 SY

 TX

RY	TX	TY	SY	SX
a	b	c	d	e

42 FS

GS

FT

GT	FS	GS	FT	GR
a	b	c	d	e

43 WP

XQ

YR

ZQ

WQ	ZR	YP	XR	ZQ
a	b	c	d	e

44 LXP

MYP

LYQ

MYQ	LXQ	MXQ	LYP	MXP
a	b	c	d	e

14

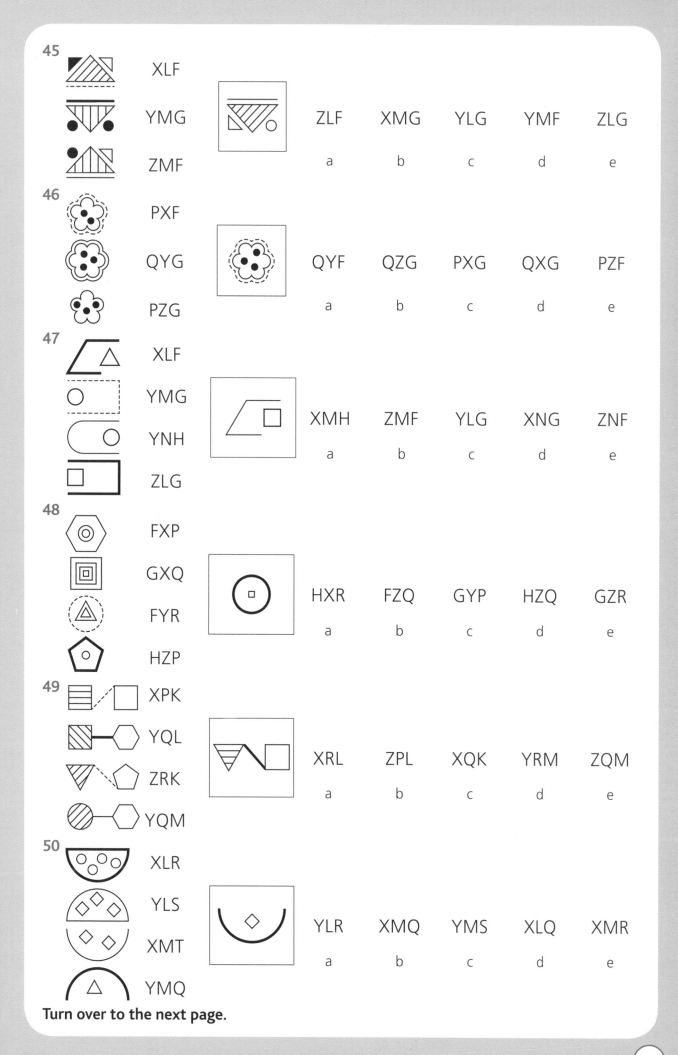

45

XLF

YMG

ZMF

	ZLF	XMG	YLG	YMF	ZLG
	a	b	c	d	e

46

PXF

QYG

PZG

	QYF	QZG	PXG	QXG	PZF
	a	b	c	d	e

47

XLF

YMG

YNH

ZLG

	XMH	ZMF	YLG	XNG	ZNF
	a	b	c	d	e

48

FXP

GXQ

FYR

HZP

	HXR	FZQ	GYP	HZQ	GZR
	a	b	c	d	e

49

XPK

YQL

ZRK

YQM

	XRL	ZPL	XQK	YRM	ZQM
	a	b	c	d	e

50

XLR

YLS

XMT

YMQ

	YLR	XMQ	YMS	XLQ	XMR
	a	b	c	d	e

Turn over to the next page.

One of the options on the right completes the pattern in the grid on the left. Circle the letter beneath the correct answer. For example:

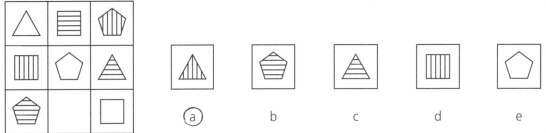

51

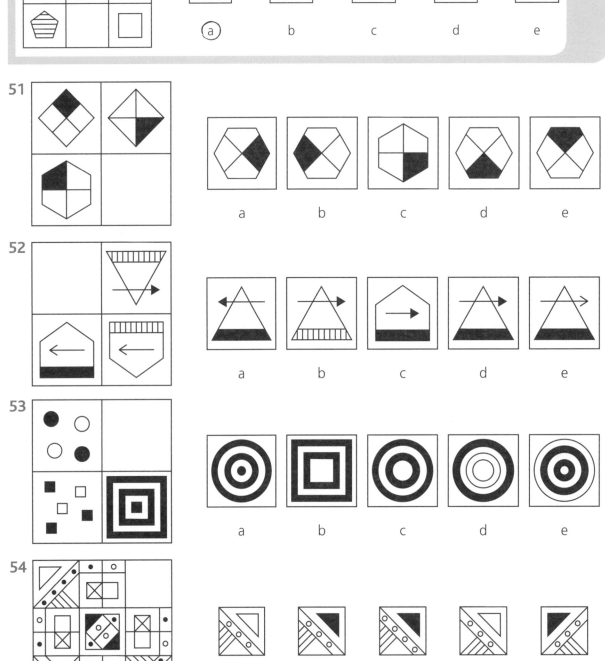

52

53

54

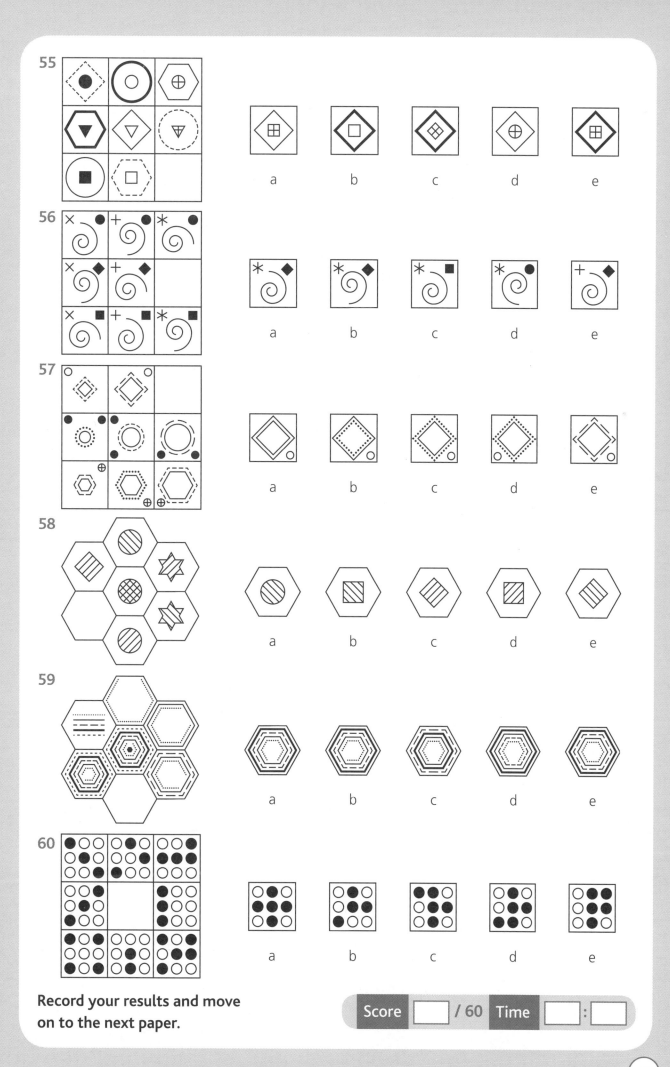

55

a b c d e

56

a b c d e

57

a b c d e

58

a b c d e

59

a b c d e

60

a b c d e

Record your results and move on to the next paper.

Score ☐ / 60 Time ☐ : ☐

Paper 10

Test time: 22:00

Look at these sets of pictures. Identify the one that is most unlike the others. Circle the letter beneath the correct answer. For example:

a   b   ⓒ   d   e

1

a   b   c   d   e

2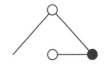

a   b   c   d   e

3

a   b   c   d   e

4

a   b   c   d   e

5

a   b   c   d   e

6

a   b   c   d   e

7

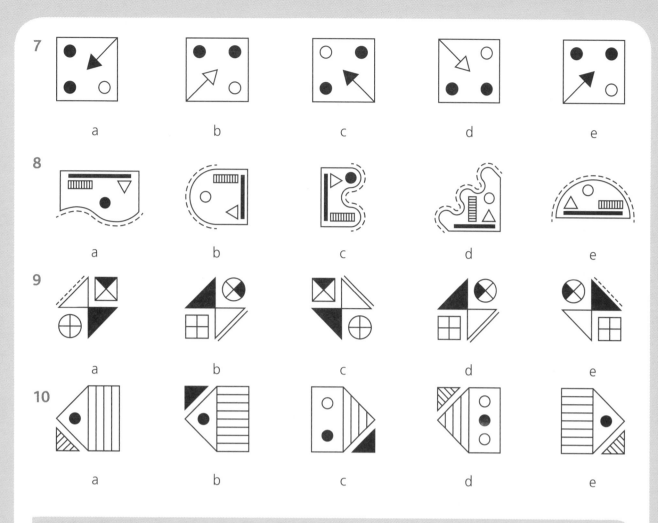

8

9

10

a   b   c   d   e

Look at the two sets of shapes on the left connected by arrows and decide how the first shapes have been changed to create the second. Now apply the same rule to the next shape and circle the letter beneath the correct answer from the five options. For example:

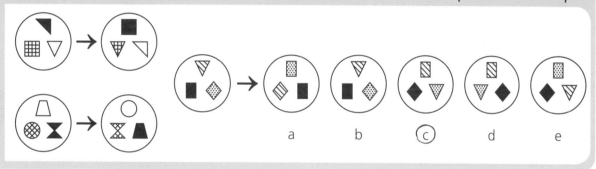

a   b   ⓒ   d   e

11

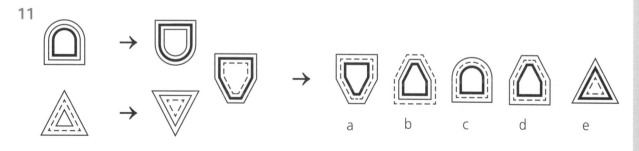

a   b   c   d   e

Turn over to the next page.

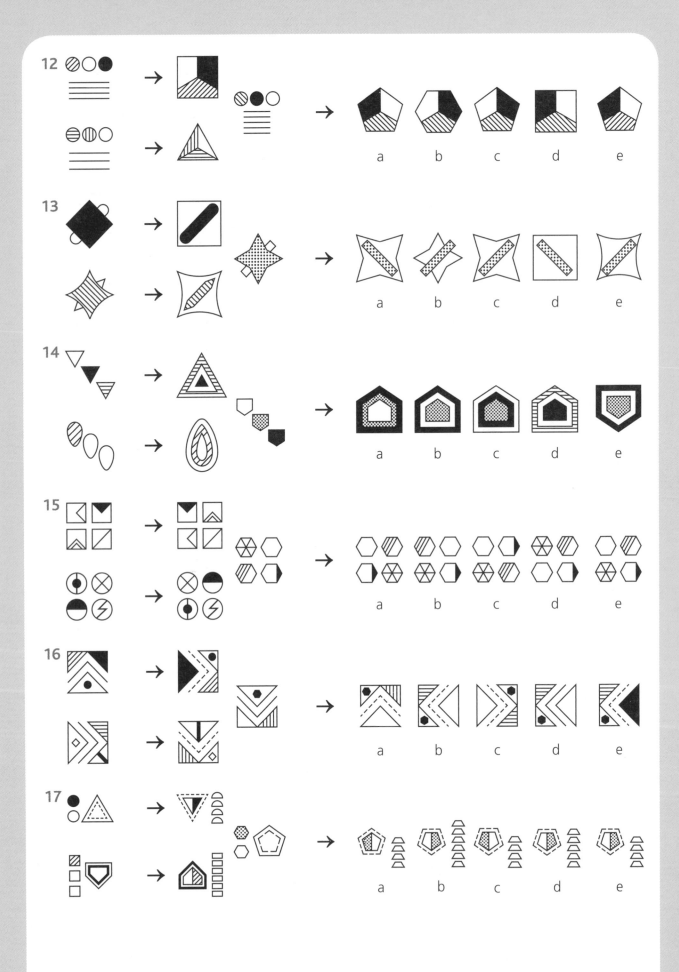

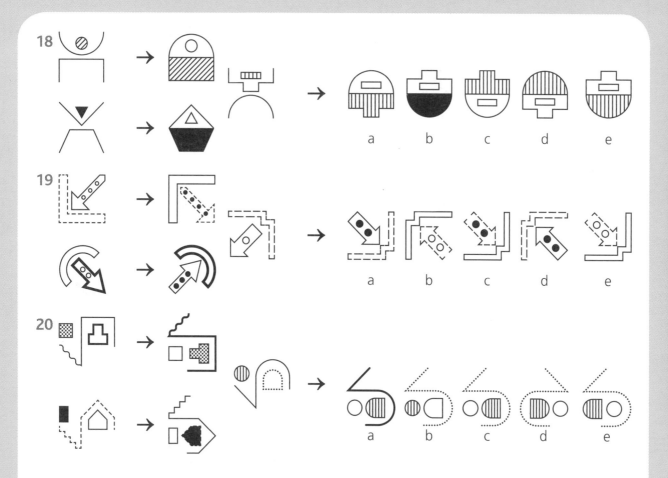

The patterns on the left are made out of paper that is black on one side and white on the other, then pasted onto a window. One of the answer options shows the same pattern viewed from the other side of the window. Notice that the colour also reverses when viewed from the other side. Circle the letter beneath the correct answer option. For example:

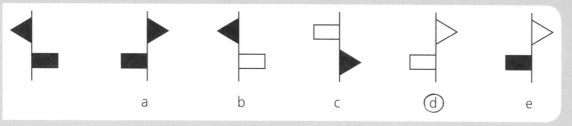

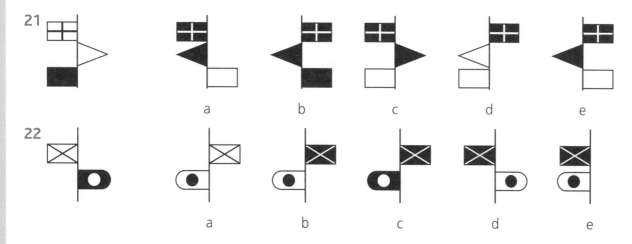

Turn over to the next page.

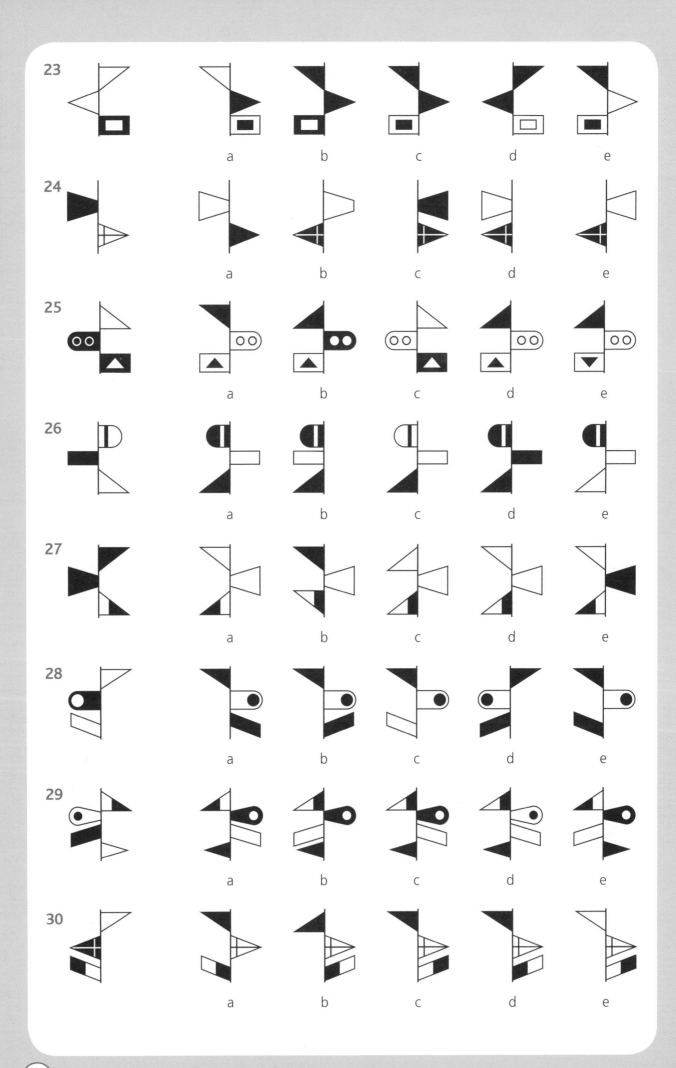

23

a b c d e

24

a b c d e

25

a b c d e

26

a b c d e

27

a b c d e

28

a b c d e

29

a b c d e

30

a b c d e

The small shape on the left can be found in one of the pictures on the right. It might be made up of one or more pieces. Circle the letter beneath the correct answer. For example:

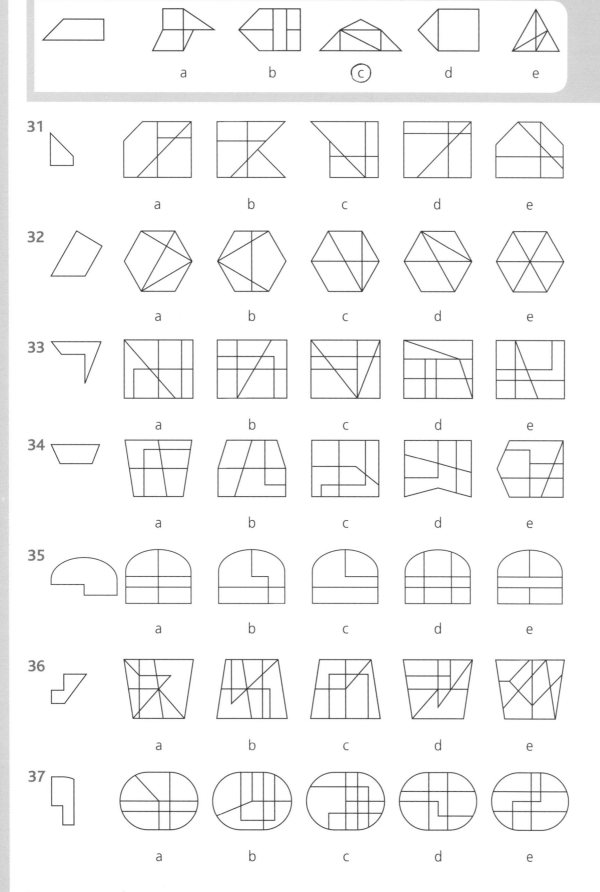

Turn over to the next page.

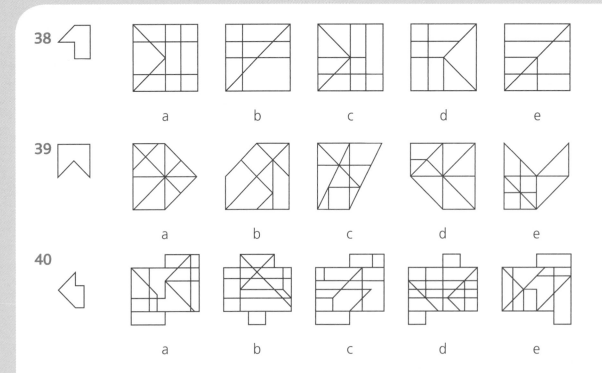

38

a   b   c   d   e

39

a   b   c   d   e

40

a   b   c   d   e

The two letters in the small boxes at the right of each large box represent a feature of the shapes in the box. Work out which feature is represented by each letter and apply the code to the box with the dashed lines. Circle the letter beneath the correct answer code. For example:

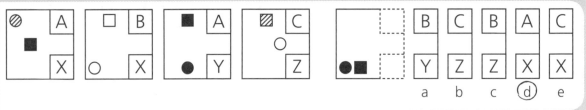

a   b   c   d   e

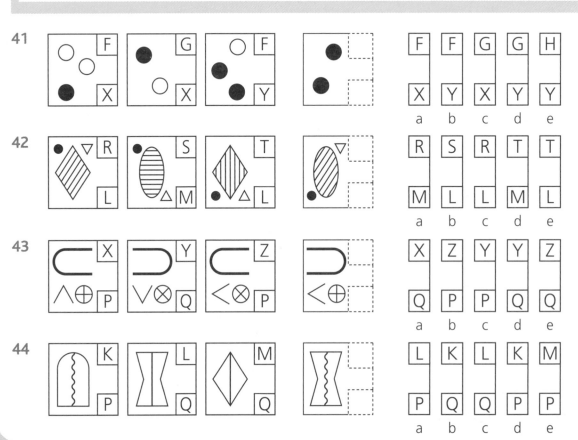

41

a   b   c   d   e

42

a   b   c   d   e

43

a   b   c   d   e

44

a   b   c   d   e

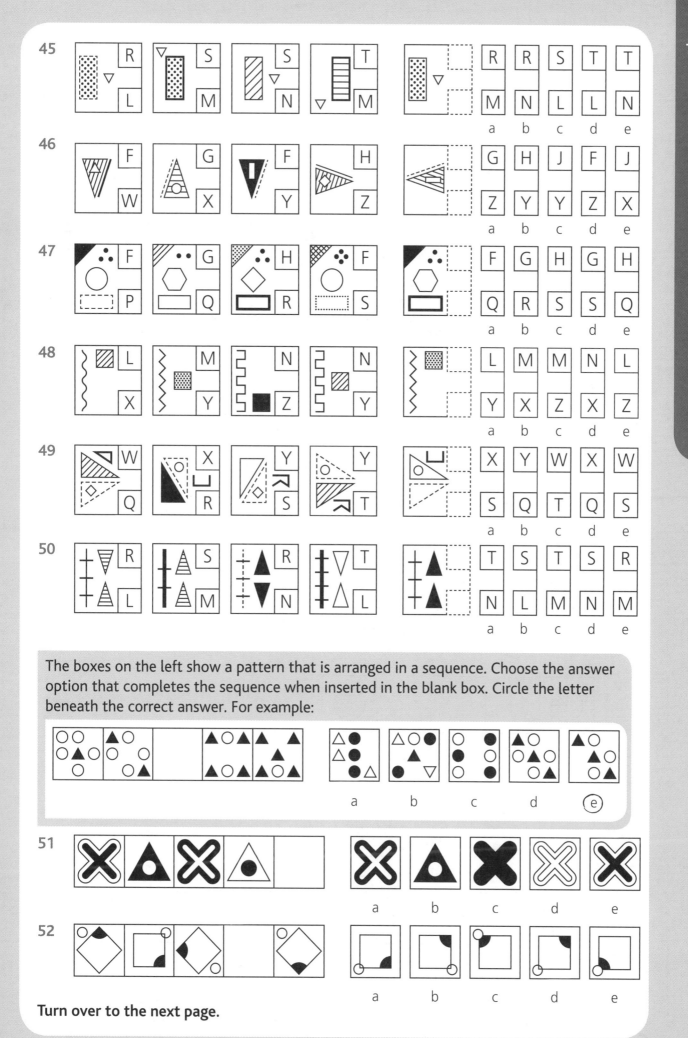

The boxes on the left show a pattern that is arranged in a sequence. Choose the answer option that completes the sequence when inserted in the blank box. Circle the letter beneath the correct answer. For example:

Turn over to the next page.

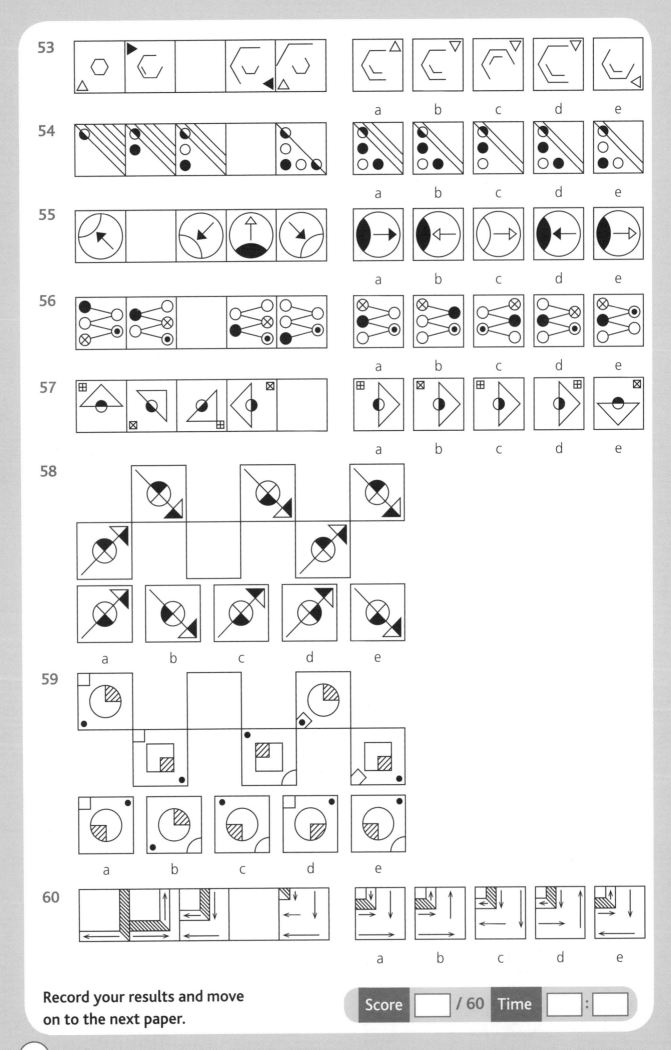

53

a b c d e

54

a b c d e

55

a b c d e

56

a b c d e

57

a b c d e

58

a b c d e

59

a b c d e

60

a b c d e

Record your results and move on to the next paper.

Score [ ] / 60 Time [ ] : [ ]

Paper 11

Test time: 11:00

Look at the shapes in the large box and decide what they have in common. Then select the option that is part of the same set. Circle the letter beneath the correct answer. For example:

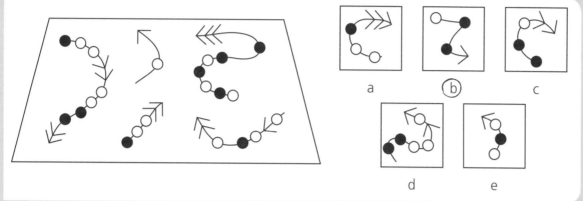

a   (b)   c

d   e

1

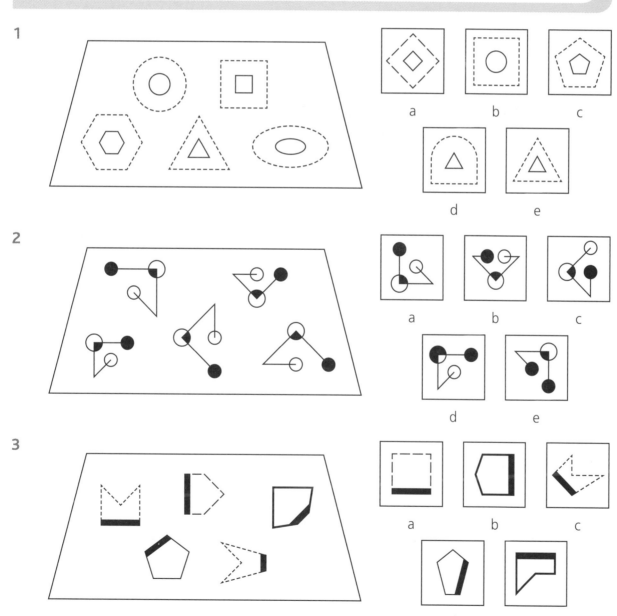

Turn over to the next page.

4

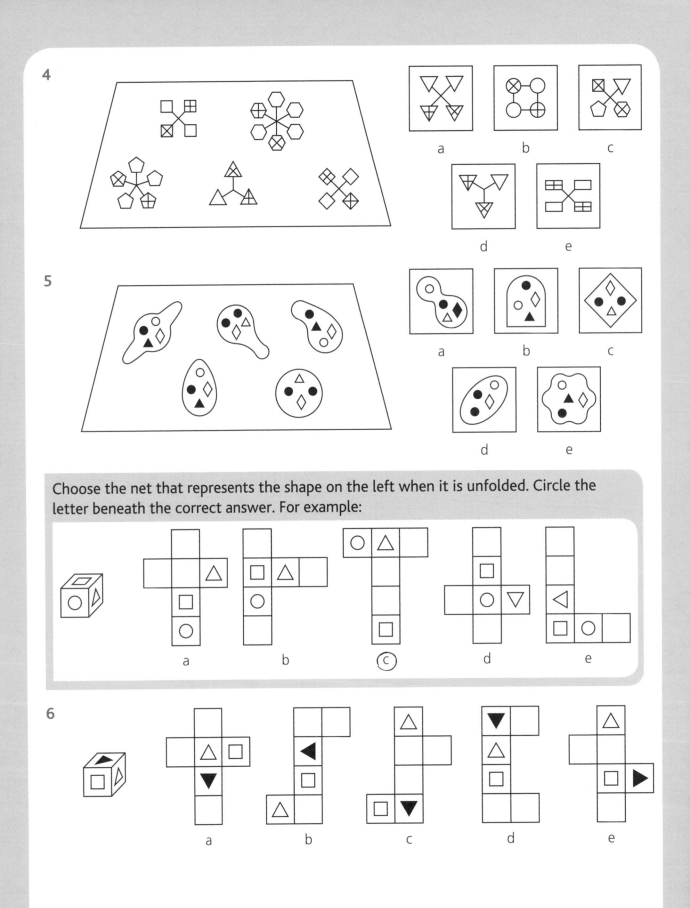

5

Choose the net that represents the shape on the left when it is unfolded. Circle the letter beneath the correct answer. For example:

a   b   ⓒ   d   e

6

a   b   c   d   e

28

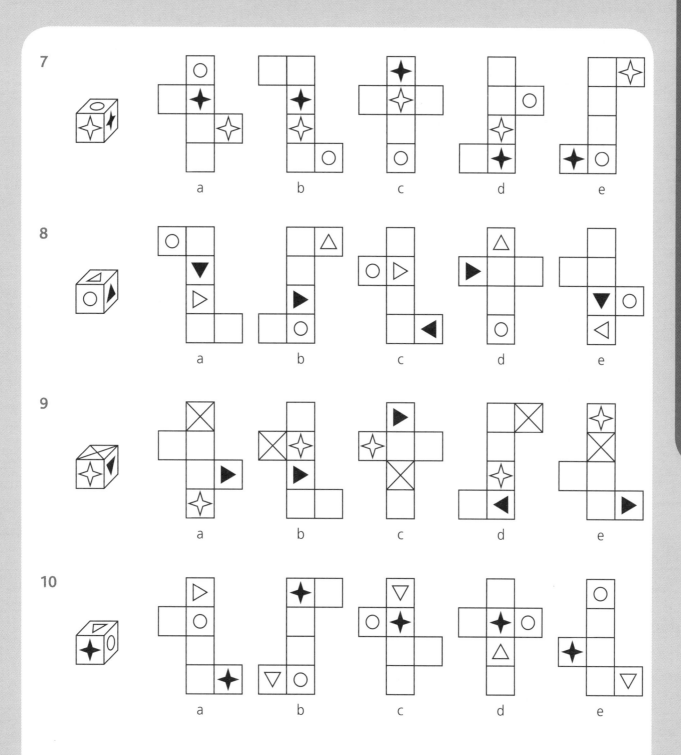

7

8

9

10

a b c d e

Turn over to the next page.

The square given at the beginning is folded in the way indicated by the arrows, and then holes are punched where shown on the final diagram. Identify the answer option that shows what the square would look like when it is unfolded. Circle the letter beneath the correct answer. For example:

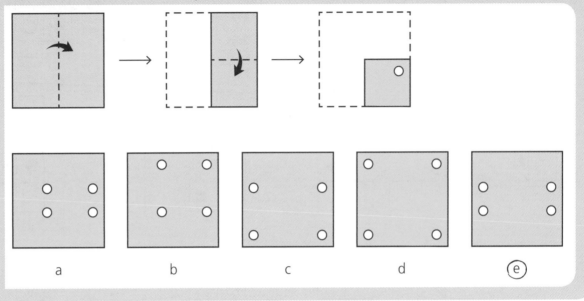

a   b   c   d   e

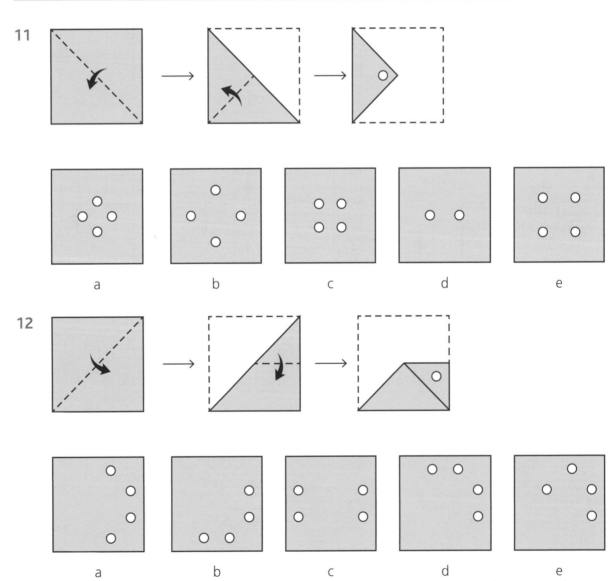

11

a   b   c   d   e

12

a   b   c   d   e

13

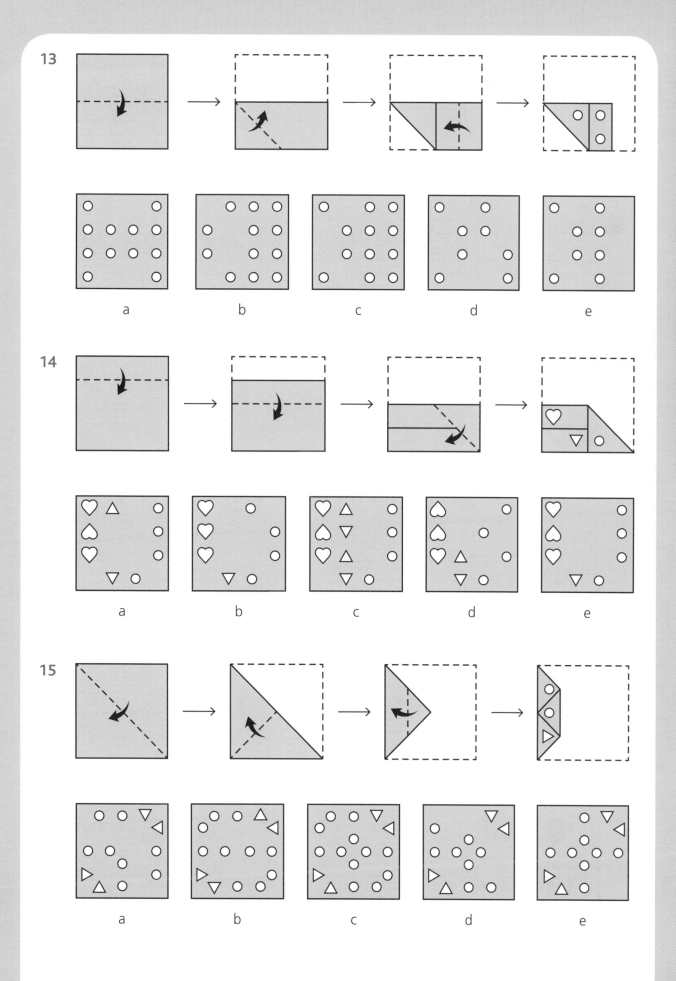

14

15

Turn over to the next page.

Each letter represents an individual feature in the picture next to it. Work out which feature is represented by each letter. Apply the code to the picture in the box and circle the letter beneath the correct answer code. For example:

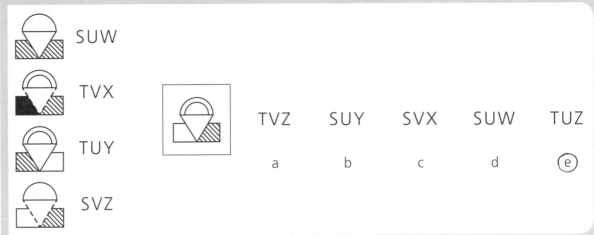

SUW

TVX

TUY

SVZ

TVZ   SUY   SVX   SUW   TUZ

a   b   c   d   (e)

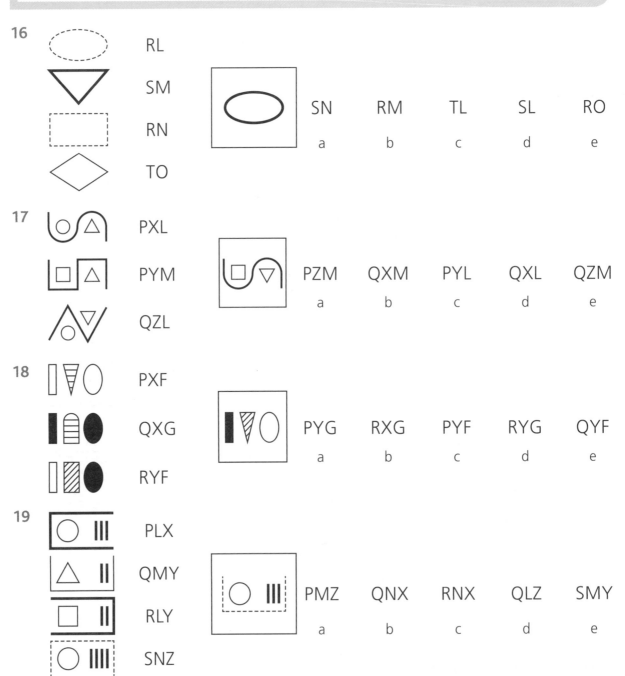

16

RL

SM

RN

TO

SN   RM   TL   SL   RO

a   b   c   d   e

17

PXL

PYM

QZL

PZM   QXM   PYL   QXL   QZM

a   b   c   d   e

18

PXF

QXG

RYF

PYG   RXG   PYF   RYG   QYF

a   b   c   d   e

19

PLX

QMY

RLY

SNZ

PMZ   QNX   RNX   QLZ   SMY

a   b   c   d   e

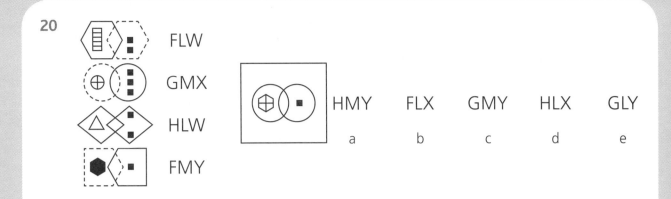

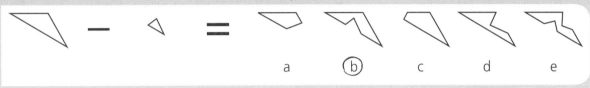

When the smaller shape on the left is removed from the larger shape before it, a new shape is made. This new shape is represented by one of the options on the right. Circle the letter beneath the correct answer. For example:

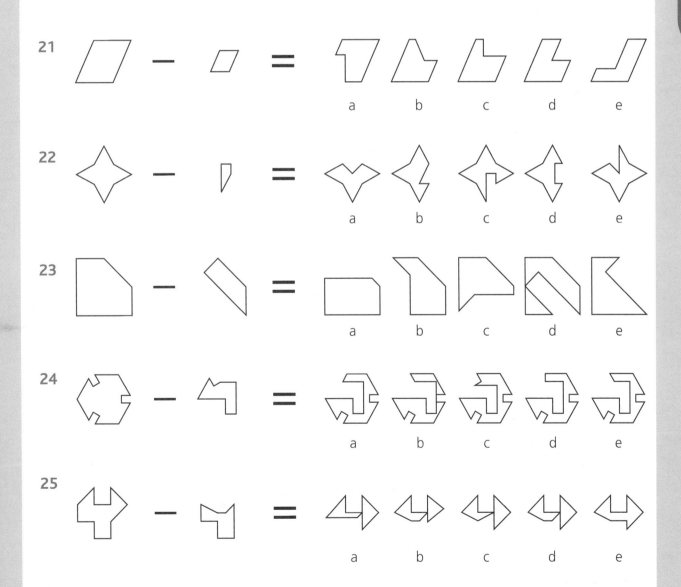

Turn over to the next page.

One of the options on the right completes the pattern in the grid on the left. Circle the letter beneath the correct answer. For example:

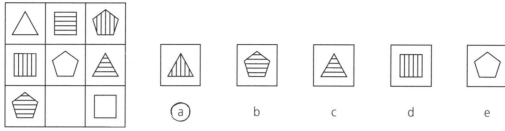

26

27

28

29

30

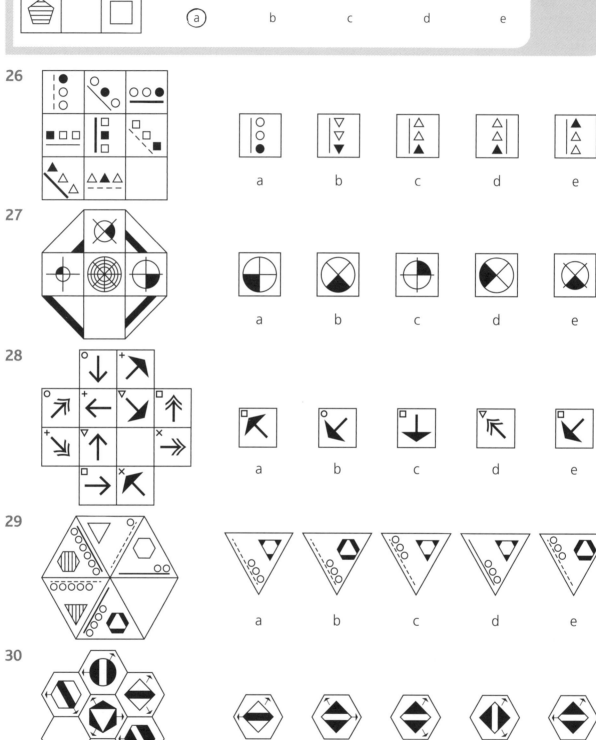

Record your results and move on to the next paper.

Paper 12

Test time: 05:00

Which of the answer options is a 2D plan of the 3D picture on the left, when viewed from above? Circle the letter beneath the correct 2D plan. For example:

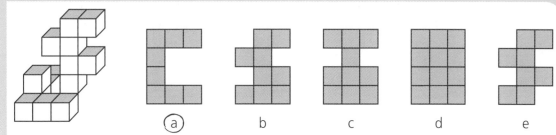

a   b   c   d   e

1

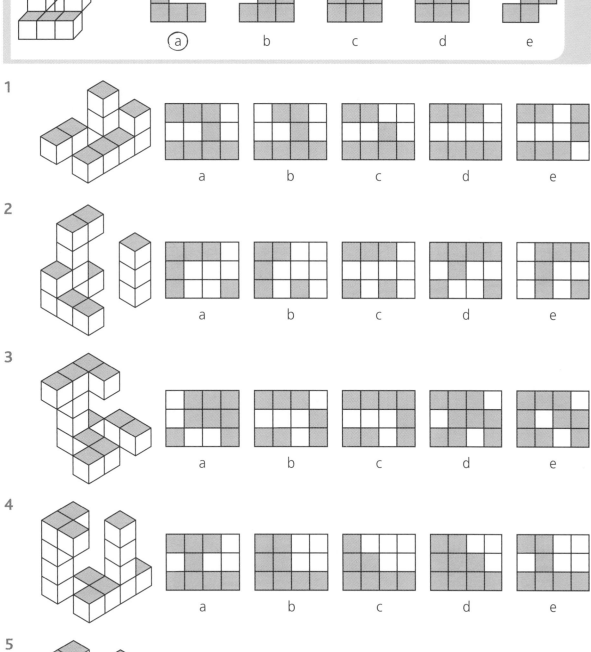

2

3

4

5

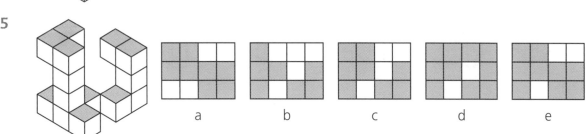

Turn over to the next page.

One group of separate blocks has been joined together to make the pattern of blocks shown on the left. Some of the blocks may have been rotated. Circle the letter beneath the blocks that make up the pattern. For example:

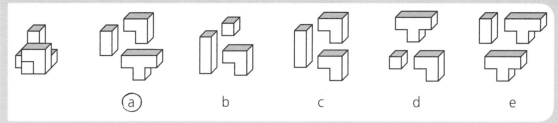

a   b   c   d   e

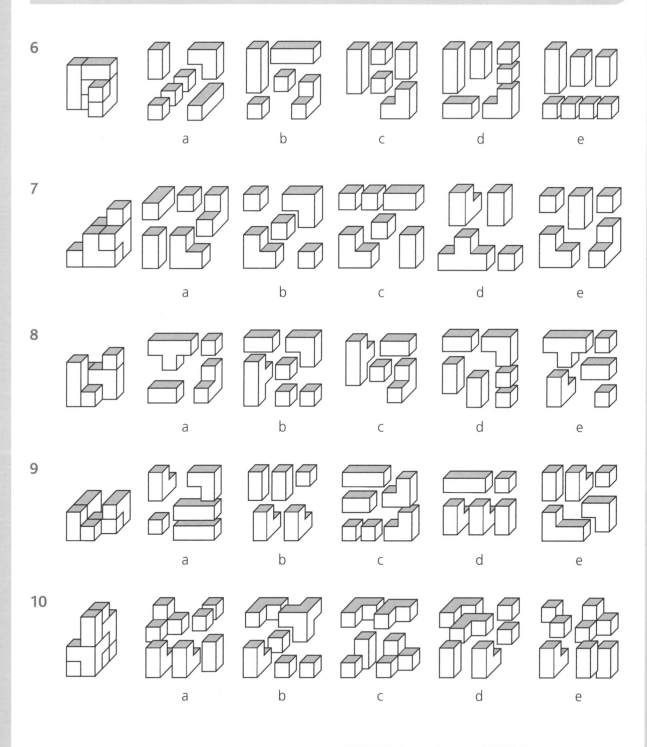

Paper 13

Test time: 22:00

Look at the first three pictures and decide what they have in common. Then select the option from the five on the right that belongs to the same set. Circle the letter beneath the correct answer. For example:

a   ⓑ   c   d   e

1

a   b   c   d   e

2

a   b   c   d   e

3

a   b   c   d   e

4

a   b   c   d   e

5

a   b   c   d   e

Turn over to the next page.

Look at the two pictures on the left connected by an arrow. Decide how the first picture has been changed to create the second. Now apply the same rule to the third picture and circle the letter beneath the correct answer. For example:

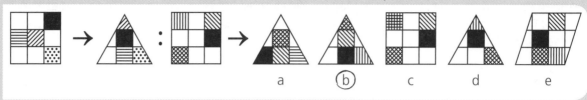

a   ⓑ   c   d   e

6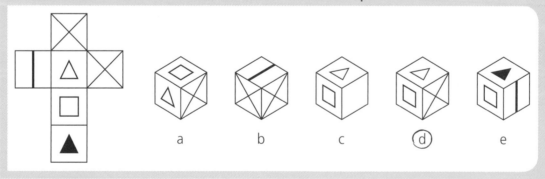

  a   b   c   d   e

7

  a   b   c   d   e

8

  a   b   c   d   e

9

  a   b   c   d   e

10

  a   b   c   d   e

Find the cube or other 3D shape that can be made from the net shown on the left. Circle the letter beneath the correct answer. For example:

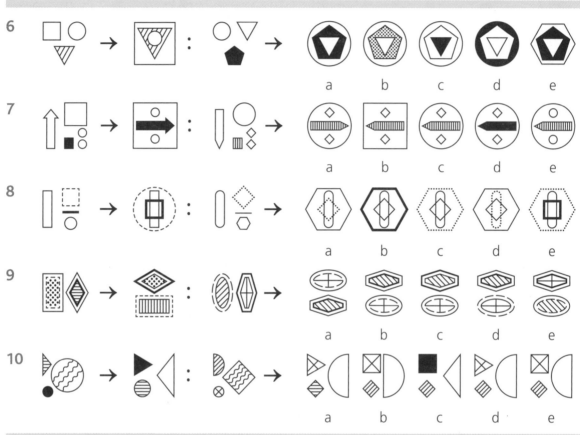

a   b   c   ⓓ   e

11

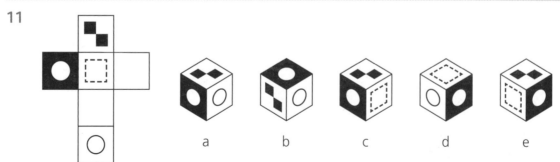

a   b   c   d   e

12

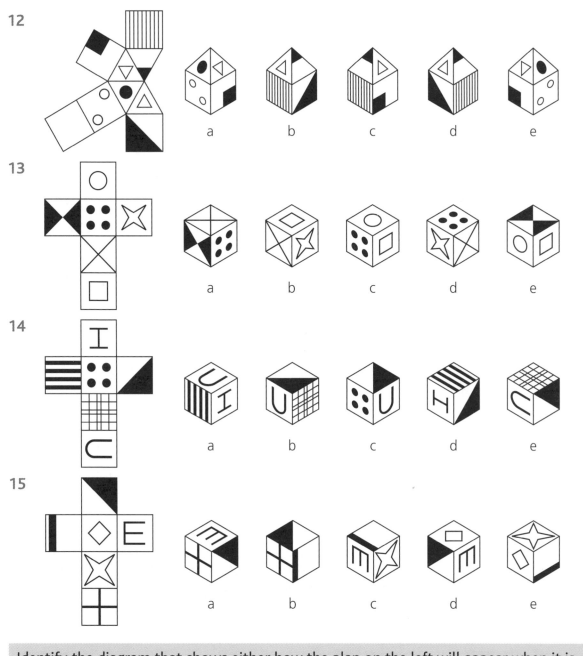

13

14

15

Identify the diagram that shows either how the plan on the left will appear when it is folded in along the dashed lines, or the plan that shows how the diagram will appear when it is folded out. Circle the letter beneath the correct answer. For example:

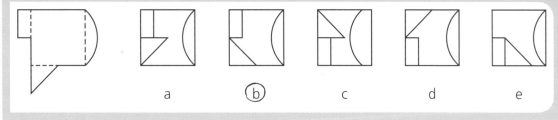

          a   ⓑ   c   d   e

16

          a   b   c   d   e

Turn over to the next page.

17

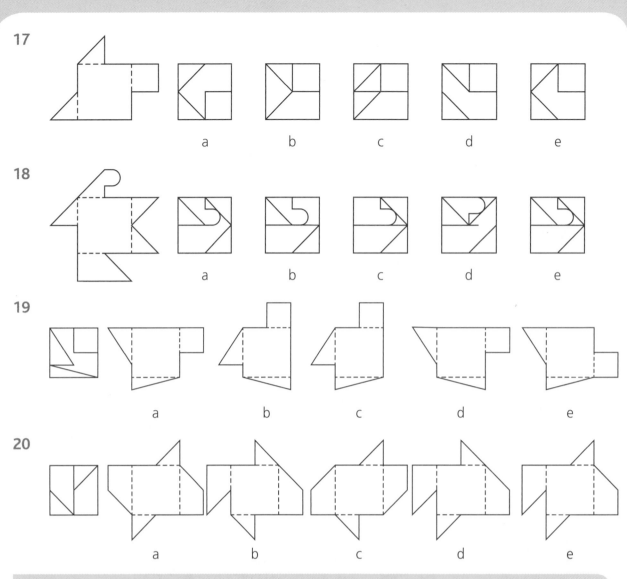

a   b   c   d   e

18

a   b   c   d   e

19

a   b   c   d   e

20

a   b   c   d   e

The picture on the left has been rotated clockwise by the number of degrees shown to give one of the pictures on the right. Circle the letter beneath the correct answer. For example:

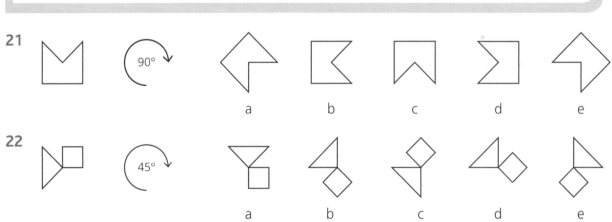

a   b   c   d   e

21

a   b   c   d   e

22

a   b   c   d   e

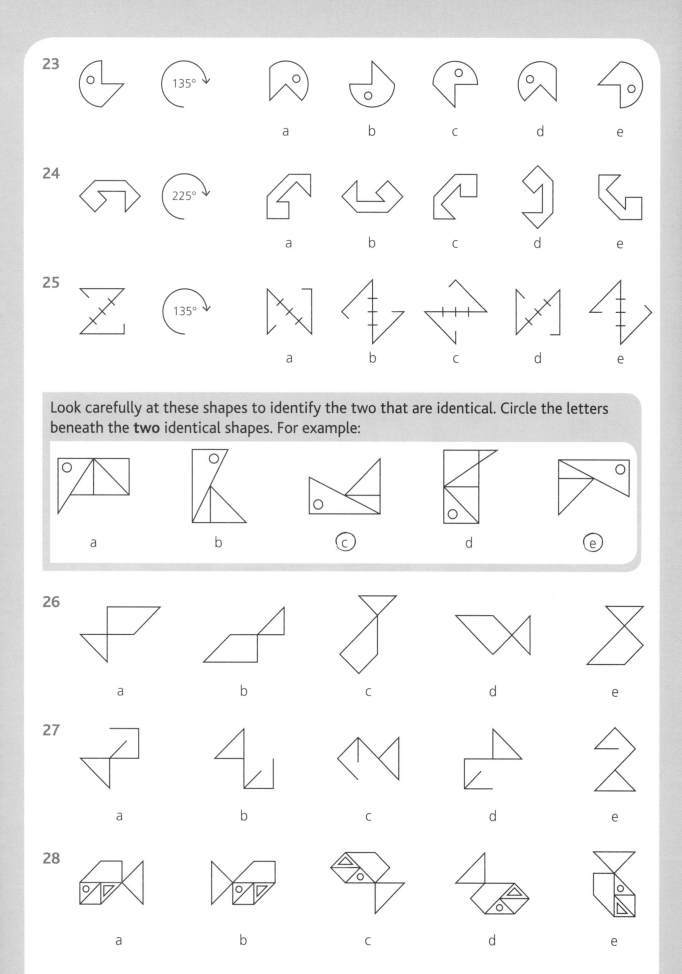

Look carefully at these shapes to identify the two that are identical. Circle the letters beneath the **two** identical shapes. For example:

a   b   c   d   e

Turn over to the next page.

29

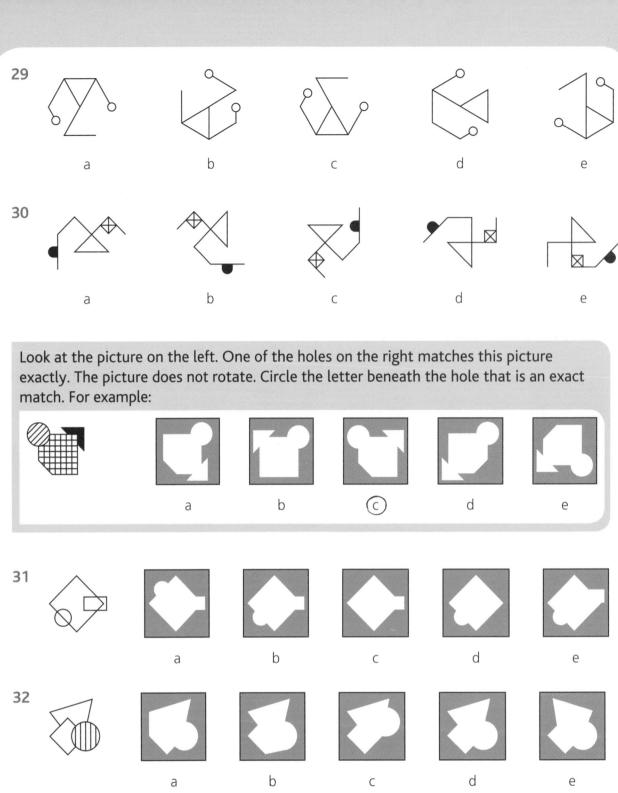

a　　　　b　　　　c　　　　d　　　　e

30

a　　　　b　　　　c　　　　d　　　　e

Look at the picture on the left. One of the holes on the right matches this picture exactly. The picture does not rotate. Circle the letter beneath the hole that is an exact match. For example:

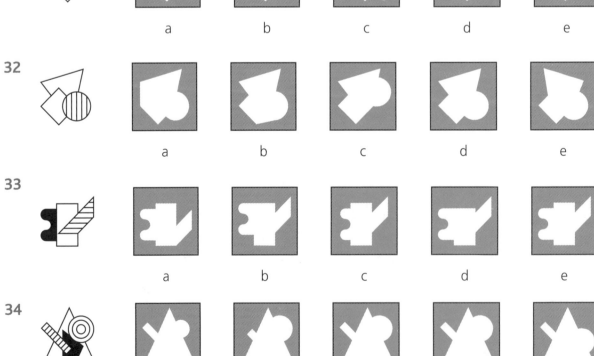

a　　　　b　　　　ⓒ　　　　d　　　　e

31

a　　　　b　　　　c　　　　d　　　　e

32

a　　　　b　　　　c　　　　d　　　　e

33

a　　　　b　　　　c　　　　d　　　　e

34

a　　　　b　　　　c　　　　d　　　　e

35

a   b   c   d   e

Two of the shapes on the right are hidden in the image on the left. Circle the **two** letters beneath the answer options that appear in the image. For example:

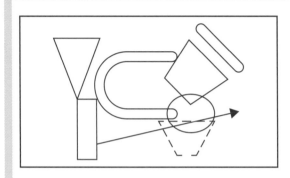

36

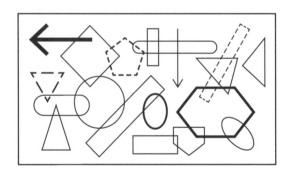

37

Turn over to the next page.

Each letter represents an individual feature in the picture next to it. Work out which feature is represented by each letter. Apply the code to the picture in the box and circle the letter beneath the correct answer code. For example:

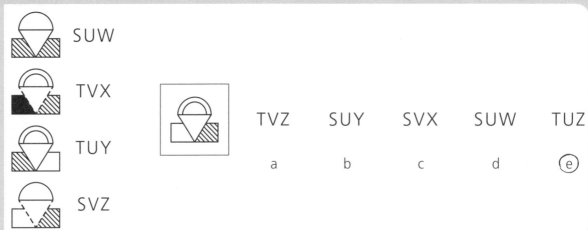

SUW

TVX

TUY

SVZ

	TVZ	SUY	SVX	SUW	TUZ
	a	b	c	d	(e)

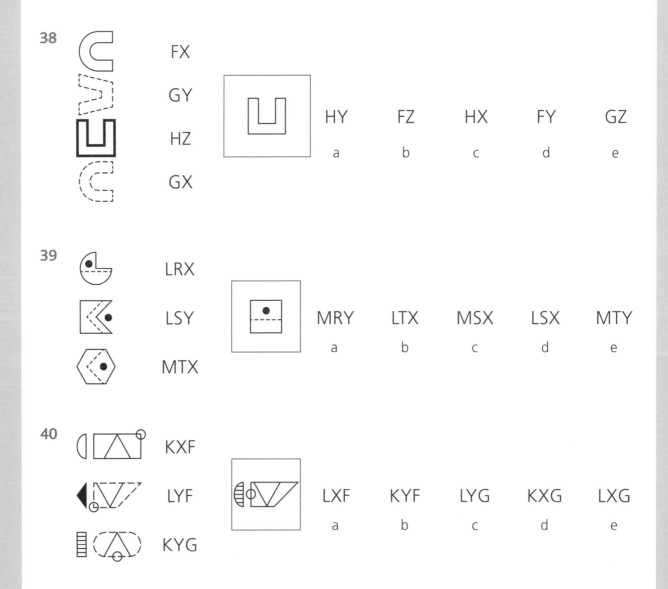

38

FX

GY

HZ

GX

	HY	FZ	HX	FY	GZ
	a	b	c	d	e

39

LRX

LSY

MTX

	MRY	LTX	MSX	LSX	MTY
	a	b	c	d	e

40

KXF

LYF

KYG

	LXF	KYF	LYG	KXG	LXG
	a	b	c	d	e

44

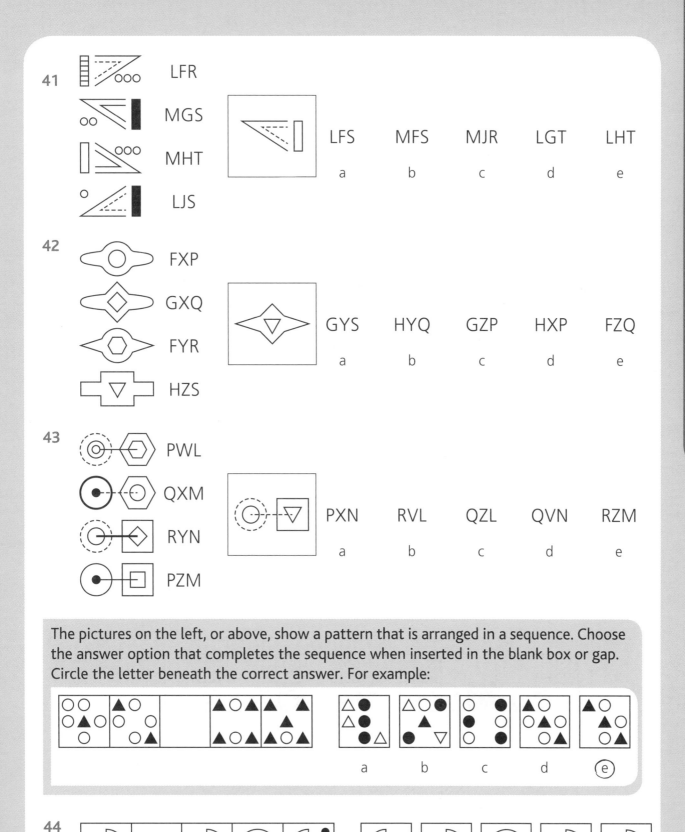

41 LFR MGS MHT LJS

LFS MFS MJR LGT LHT
 a   b   c   d   e

42 FXP GXQ FYR HZS

GYS HYQ GZP HXP FZQ
 a   b   c   d   e

43 PWL QXM RYN PZM

PXN RVL QZL QVN RZM
 a   b   c   d   e

The pictures on the left, or above, show a pattern that is arranged in a sequence. Choose the answer option that completes the sequence when inserted in the blank box or gap. Circle the letter beneath the correct answer. For example:

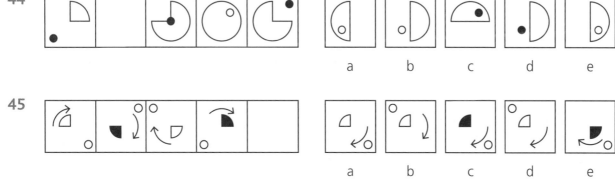

 a   b   c   d   e

44

 a   b   c   d   e

45

 a   b   c   d   e

Turn over to the next page.

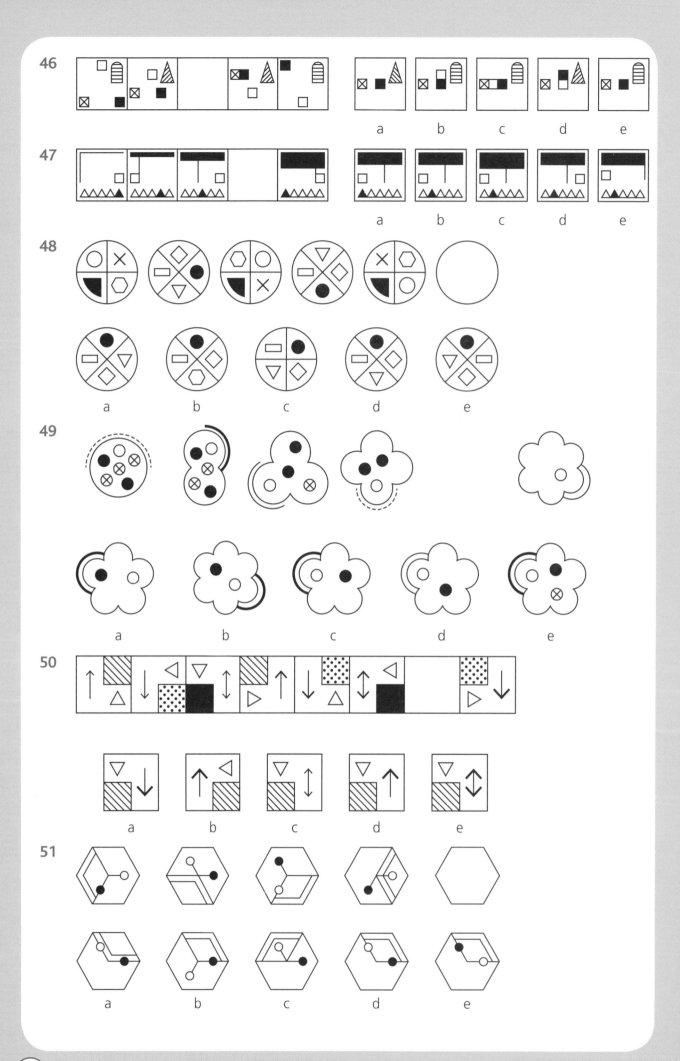

52

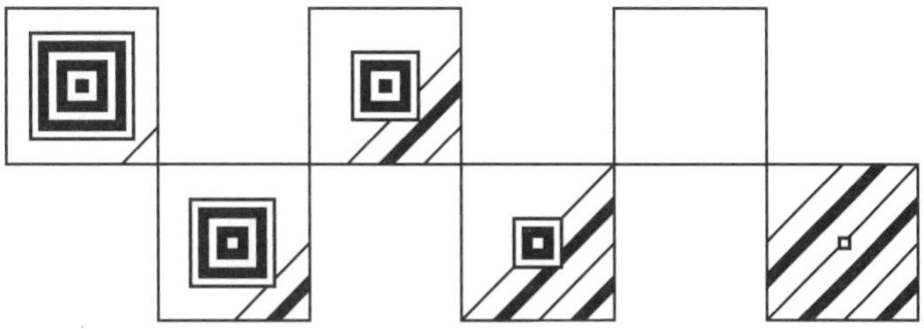

a   b   c   d   e

One of the options on the right completes the pattern in the grid on the left. Circle the letter beneath the correct answer. For example:

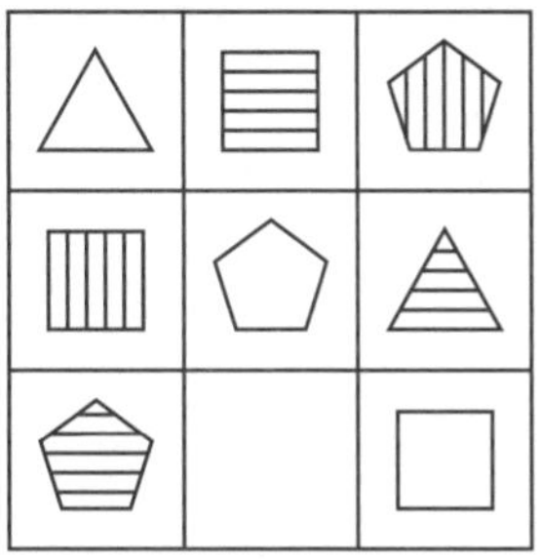

a   b   c   d   e

53

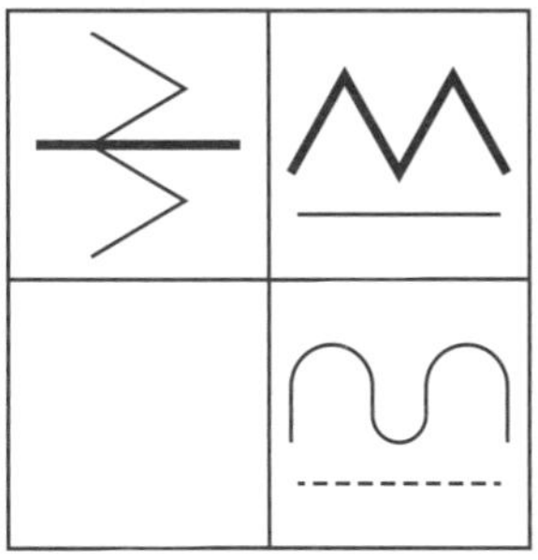

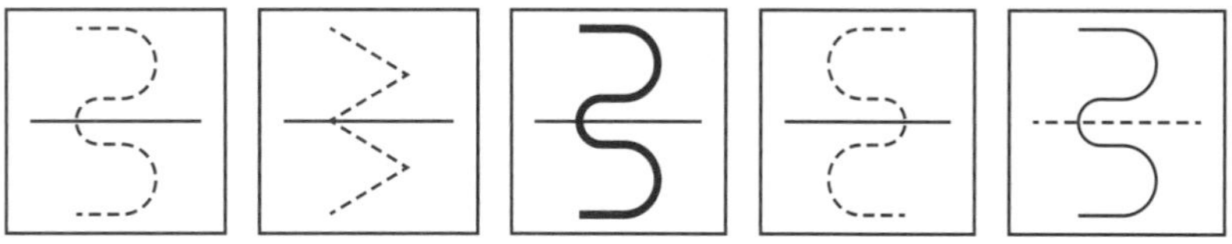

a   b   c   d   e

54

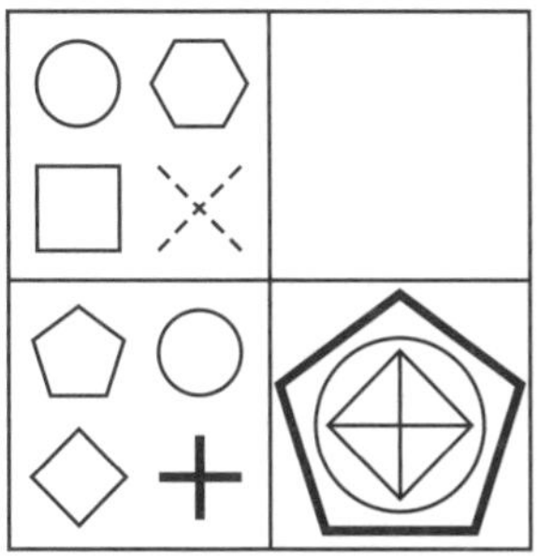

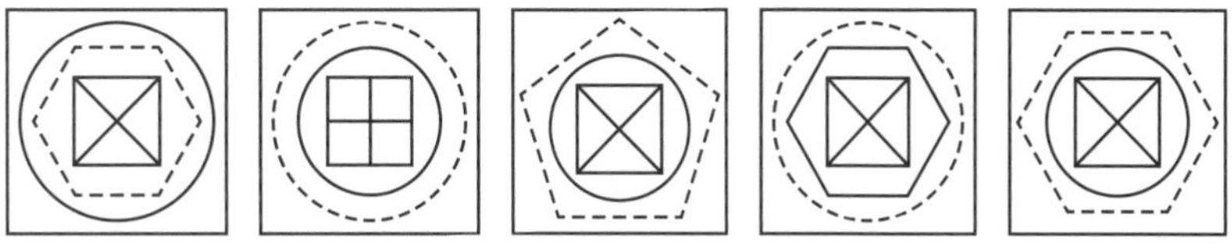

a   b   c   d   e

55

a   b   c   d   e

Turn over to the next page.

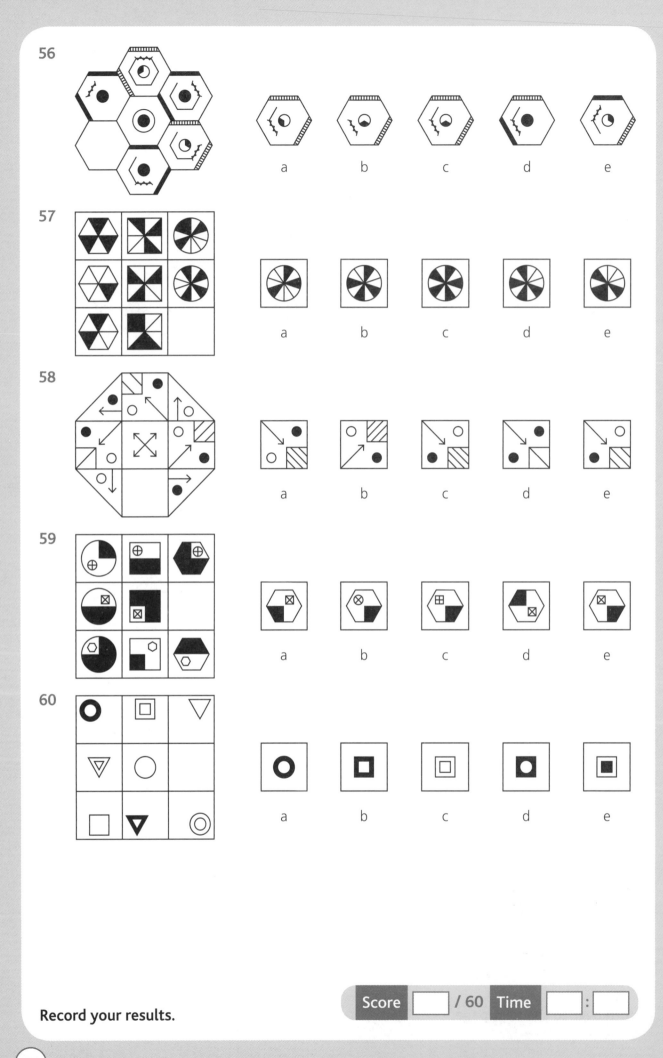

56

57

58

59

60

Record your results.

Score [   ] / 60   Time [   ] : [   ]

Answers

Please note all questions are worth one mark.

Paper 9 (page 9)

1 **e** **position** – each picture contains a circle inside/outside the large circle and one overlapping the edge; **shading** – (a) the circle inside/outside the circle is always black, (b) the overlapping circle is black on the portion inside the circle if the other circle is inside, and black on the portion outside the circle if the other circle is outside.

2 **a** **number** – each picture includes a six-sided shape; **position** – the circle is always inside the large shape.
Distractors: **position** – the position of the cross is a distractor – it can be inside or outside the shape.

3 **c** **symmetry** – each picture has *one* line of symmetry (note: option **e** has two so is incorrect).
Distractors: **number** – (a) the number of shapes is unimportant, (b) the number of sides of the shapes individually or added together is unimportant; **shape** – the type of shape is unimportant.

4 **d** **position** – the circle is always one and two spaces away from the other two small shapes (one space away from one; two spaces away from the other). **shading** – either the rectangle or the triangle is shaded.
Distractors: **number** – the number of bulges in the 'cloud' shape is always the same (7); **position** – the position of the shapes around the edge isn't important, apart from the relationship described above.

5 **c** **shading** – the shaded semi-circle and side always take the same shading; **position** – (a) the shading of the semi-circle is always opposite the shaded side, and the other side of the central line, (b) the small circle overlapping the central line is always to the right (when looking at the figure with the shading beneath the line), (c) the other circle is two vertices away from this on the shaded side of the hexagon.

6 **b** **proportion** – the solid line around the inner shape is drawn around one half of the shape; **shape** – there is one white triangle outside the large shape.
Distractors: **shading** – the shading in the pictures is not important; **shape** – the type of shapes are unimportant.

7 **c** **number** – (a) there are either eight sides or shapes with one side, (b) there is always one black circle in addition to the shapes.
Distractors: **shading** – the shading on the shapes other than the black circle is not important; **proportion** – the size of the shapes is not important.

8 **a** **shape** – each picture has two lines with a triangle at the end of one line and a cross at the end of the other; **position** – the small black circle is always inside the small 'loop' shape and the white circle is always above the loop.
Distractors: **shape** – the triangle shown is always a right-angled triangle; **position** – (a) the triangle is positioned randomly at the end of one line, (b) the cross always fits at the end of the other line, (c) it doesn't matter which line the loop is on as long as it follows the rules above.

9 **d** **proportion** – the two lines in each picture are exactly the same length.
Distractors: **shape** – (a) the shape of the shaded figure is not relevant, (b) the shape of the line is not relevant; **direction** – the direction of the shading is unimportant; **position** – the position of the shapes in relation to each other is unimportant; **line style** – as all the answers have the same line style this can also be discounted.

10 **b** **shape** – (a) the large shape always has at least one *horizontal* straight edge, with the shortest edge open along half of the base, (b) there is always a quadrilateral inside one section of the enclosed side; **position** – the position of the line dividing the enclosed section of the shape runs between the long edges of the shape.
Distractors: **number** – the number of short lines is unimportant; position – the position of the quadrilateral – either in the upper or lower section of the closed shape is unimportant.

11 **c** **reflection** – the left-hand shape reflects in a vertical line; **shading** – the shading from the right-hand shape becomes the shading in the reflected shape.

12 **d** **proportion** – the outer shape loses the left half, **rotation** – the inner shape rotates 45°.
Distractor: **number** – there is no change in the number of sides of the inner figure.

13 **d** **proportion/translation** – top and bottom shapes enlarge as the shapes move together to sit one on top of the other, (a) the bottom shape becomes medium-sized, (b) the top shape becomes the largest size.

14 **e** **number** – the number of small lines represents the number of sides in the bottom shape in the second picture; **translation** – the shape on the left becomes the upper shape; **shading** – (a) the shading of the bottom-right shape becomes the shading on the top shape, (b) the shading of the top-right shape becomes the shading/pattern of the bottom shape.

15 **a** **reflection** – the outer shape reflects in a horizontal line; **proportion** – the large shape becomes the smaller, the small shape become the larger; **line style** – the line style of the reflected shape becomes thick and solid (the line style of the other shape is unaltered).

16 **d** **proportion/translation** – the shapes all change in proportion and sit on top of each other, (a) the shapes in the diagonal top left to bottom right become the largest and third-largest shapes, (b) the shapes in the diagonal top right to bottom left become the second largest and smallest shapes; **line style** – the line style of the two sets of shapes swaps.

17 e **rotation** – the large shape rotates 45°; **position** – (a) the top and bottom-left shapes reflect in a central horizontal line and move to the centre top and bottom of the picture, (b) the shape on the top right moves into the middle of the picture; **shading** – (a) the top-left and bottom-left shapes swap shading, (b) the pattern/shading from the bottom-right shape moves into the central shape and this bottom shape disappears.

18 b **rotation** – the picture rotates 180°; **shading** – the shading in the top and bottom panels remains in their original positions; **position** – the small squares scatter through the large shape (although the number of squares doesn't change).

19 d **proportion** – (a) the small shape in the top left of the larger shape enlarges to become the larger shape, (b) the top half of the larger shape reduces in size to fit in the centre; **rotation** – this shape rotates 45°; **line style** – (a) the line style of the outer shape is taken from the bottom small shape, (b) the inner shape takes its line style from the top small shape (or outer shape – they are both the same).

20 c **rotation** – the large shape rotates 90°; **line style** – the double lines that create this shape swap style; **position** – (a) the right-hand shape moves outside the larger shape (and doubles); (b) the other two shapes move inside the shape with one of them inverting; **shading** – the left-hand shape gives its shading to the outer shapes, (b) the middle shape gives its shading to the upper internal shape with the bottom shape taking its shading from the right-hand shape.

21 e The grid on the top face of the cube has moved to the left on the second cube, so we know the cube has rolled to the left. Rotating the second cube in the same way will have no effect on the circle on the front face; the triangle on the top face will move to the left-hand face and continue to point towards the front of the cube.

22 c The arrow on the top face has moved to the face at the front, so we know the cube has rolled forward towards the viewer. Rotating the second cube in the same way will mean the stripe on the top rolls forward so that the parallel lines run vertically down the front face; the black arrow in the left will rotate forwards 90°.

23 e The circles on the top face disappear so must roll out of sight, so you need to look at the other faces to see what is happening; the triangle on the left-hand face ends up on the top, so the cube must have rolled to the right. Rotating the second cube in the same way will mean that the diagonal on the left-hand face will run bottom left to top right in the upper face (and the black section will be on the left). Next, look at the triangle on the front face – it is pointing upwards on the first cube, so rotating it 90° clockwise will mean it points to the right on the second cube.

24 d The rectangle on the top face has moved to the face at the front, so we know the cube has rolled

forward towards the viewer. Rotating the second cube in the same way will mean the triangle on the top face will be on the face at the front, pointing downwards from the top of the face. If the arrow on the left-hand face is also rotated 90° towards the viewer it will be pointing downwards in the second cube.

25 c All the faces on the first cube appear on the second cube but have moved one place anticlockwise around the cube, so it must have been rolled to the left. Apply these changes to the second cube in the same way to arrive at the answer.

26 b The arrow on the front face has moved to the top face so the cube has rolled away from the viewer. However, as the arrow has changed direction too, the cube must also have been rotated anticlockwise. Apply these changes one at a time to the second cube: rolling the cube away from the viewer will move the triangle to the top face pointing towards the front, the square on the top face will disappear from sight and the line on the left-hand face will remain on this face, but pointing bottom left to top right. If the cube is then rotated 90° anticlockwise, the triangle will be on the top face, pointing to the right, the diagonal line will be on the front face and run top left to bottom right, finally the square will appear again on the left-hand face at the back in the bottom corner.

27 d The parallel lines on the top face have moved to the front face so the cube has rolled towards the viewer. However, as the lines run down rather than across, it must have been rotated through 90° clockwise (because the face with the circle is still visible). Apply these changes one at a time to the second cube: rolling the cube towards the viewer will move the white triangle to the front face and pointing towards the left, the striped triangle will be beneath and the black triangle will remain on the left-hand face but move to the back face to point towards the front. If the cube is then rotated 90° clockwise, the white triangle will move to the bottom of the front face, the striped triangle will appear on the left-hand face and point from front to back. Finally the black triangle will be on the top face pointing towards the front.

28 c The arrow on the front face shows that the figure just makes one movement and rotates 90° clockwise. Applying this change to the second cube moves the triangle on the left-hand face to the top face – this triangle will continue to point to the left. The circle on the front when rotated will remain unchanged (option **a** is a distractor showing a 90° rotation).

29 e Following the '7' shape on the front face shows the cube has been rotated away from the viewer. However, as it has changed direction as well, the cube has also been rotated anticlockwise. Apply these changes one at a time to the second cube: rolling the cube backwards will put the circle and

diagonal line on the top face with the diagonal continuing to run top left to bottom right. The black triangle will disappear from view and the parallel lines on the left-hand face will move to the base of this face. When the cube is then rotated anti-clockwise, the diagonal on the top face will change direction to run top right to bottom left, the parallel lines will be at the bottom of the front face and the black triangle will appear on the left-hand face in the top left-hand corner.

30 **e** Following the arrow on the front face shows the cube has been rotated away from the viewer. However, as it has changed direction as well, the cube has also been rotated anticlockwise. Apply these changes one at a time to the second cube: rolling the cube backwards will put the 'Z' shape on the top face, the triangle will disappear from view and the rectangle will be attached to the base of the left-hand face. When the cube is then rotated anti-clockwise, the 'Z' shape will look like an 'N' on the top face, the rectangle will move to the front face, remaining at the bottom and the black triangle will appear on the left-hand face, pointing towards the front of the cube.

31 **a, b, c** **a** sits beneath shape **c** to square up the bottom of shape, **b** sits beneath **a** to match the picture shown on the left.

32 **b, d, e** **e** sits on the top left of **d**. **b** then rotates 90° clockwise to sit on the diagonal at the right of this shape. Finally, the whole shape rotates a further 90° clockwise to match the picture shown on the left.

33 **a, c, d** **c** rotates 180° and sits on the top of **a**, **d** then slots directly into the top of this figure to complete the rectangle. Finally, the whole shape rotates 90° to match the picture shown on the left.

34 **b, d, e** **b** fits into the shape at the top of **e**, forming two sides of the final picture. **d** then slots directly into the bottom to complete the hexagon.

35 **a, c, e** **a** rotates 90° and forms the right-hand side of the picture, **e** then slots directly into the top with **c** rotating 180° to fit into the base of the shape.

36 **b, c, e** **b** rotates 90° anticlockwise and forms the base of the picture, **c** then rotates 180° to fit into the diagonal at the top of **b**. Finally, **e** rotates 180° to fit beneath **c** to complete the picture.

37 **a, c, d** **a** rotates 90° clockwise to form the right-hand side of the picture, **c** rotates 90° anticlockwise to form the top left of the picture. Finally **d** slots in at the base, beneath **c**, to complete the picture.

38 **b, c, e** **b** rotates 45° anticlockwise to sit beneath **e**, forming the bottom of the hexagon, **c** then rotates 135° anticlockwise to sit on top of **e** and form the top right of the hexagon.

39 **a, c, e** **a** rotates 135° clockwise to form the top of the picture, **c** then sits beneath this

at the base, on the left. Finally **e** rotates 90° anticlockwise to sit on the right of **c** to complete the picture.

40 **b, c, d** **a** forms the top right-hand side of the picture, **c** then rotates 45° anticlockwise to sit beneath this at the base. Finally **d** rotates 135° anticlockwise to sit on the left of **b** to complete the picture.

41 **c** **shape** – the first letter represents the shape: R is a rectangle, S a diamond and T an oval; **shading** – the second letter represents the shading of this shape: X diagonal stripes and Y vertical stripes.

42 **a** **shape** – the first letter represents the small shape: F is a circle and G a triangle; **line style** – the second letter represents line style of the large shape: S is dashed and T is solid. Distractor: **direction** – the direction the triangle 'points' in is unimportant.

43 **b** **shape** – (a) the first letter represents the second shape: W is a circle, X a right-angled triangle, Y an equilateral triangle and Z a shield, (b) the second letter represents the first shape: P an equilateral triangle, Q a right-angled triangle, R an oval. Distractor: **shading** – the shading is unimportant.

44 **c** **line style** – (a) the first letter represents the top line: L is straight, M is zig-zag, (b) the second letter represents the bottom line: X is curved wavy pattern, Y has a square-shaped pattern, (c) the third letter represents the middle line: P is a wavy pattern and Q a zig-zag.

45 **e** **line style** – the first letter represents line style of the line outside the triangle: X is dashed, Y is solid and thick, Z is solid and thin; **shading** – the second letter stands for the shading of the large triangle: L diagonal stripes, M vertical stripes; **direction** – the third letter represents the direction of the triangle: F 'pointing' upwards, G downwards. Distractors: **shape** – the small shapes are not related to the code; **shading** – the shading of the small shapes is also not related to the code.

46 **d** **number (1)** – the first letter represents the number of 'petals' on the flower shape: P is five, Q is six; **line style** – the second letter represents the line style of the outer 'petals': X is dashed, Y is solid and Z is none; **number (2)** – the third letter represents the number of circles inside the flower shape: F is two, G is three.

47 **e** **shape (1)** – the first letter represents the small shape: X is a triangle, Y a circle and Z a square; **line style** – the second letter represents the line style of the outer shape: L is thick and solid, M is dashed, N is thin and solid; **shape (2)** – the third letter represents the outer shape: F is the shape with the pointed end, G the square end and H the rounded end. Distractors: **position** – the position of the small shapes doesn't relate to the code; **direction** – the direction the large shape is 'pointing' in doesn't relate to the code.

48 **d** **number** – the first letter represents the number of concentric shapes: F is two, G is three, H is

one; **line style** – the second letter represents the line style of the outer shape: X is thin and solid, Y is dashed, Z is thick and solid; **shape** – the third letter represents the inner shape: P is a circle, Q a square, R a triangle.
Distractor: **shape** – the outer shape does not relate to the code.

49 a **shape** –the first letter represents the shape on the right: X is a square, Y a hexagon, Z a pentagon; **position** – the second letter represents the position of the central line: P bottom of the left-hand shape to top of the right, Q horizontal, R top of the left-hand shape to bottom of the right; **line style** – the third letter represents the style of the central line: K is dashed, L is solid and thick, M is solid and thin.
Distractors: **shape** – the left-hand shape does not relate to the code; **shading** – the shading of this shape does not relate to the code.

50 b **position** –the first letter represents the position of the semi-circle: X is with the curved side at the bottom, Y with the curved side on the top; **shape** – the second letter represents whether there is a straight line at the top of the semi-circle: L stands for a line present, M for a line absent; **number** – the third letter represents the number of small shapes: Q is one, R is four, S is three, T is two.
Distractors: **shape** – type of small shape doesn't relate to the code; **line style** – the line styles of the semi-circle and straight line don't relate to the code.

51 a **shape** – the shape changes on each row; **rotation** – the cross in the shape rotates 45° between the left and right columns; **position** – the shading moves one quarter clockwise between the left and right columns.

52 d **shape** – (a) the shape changes on each row, (b) the shape of the arrowhead changes between rows; **direction** – the direction of the arrow changes between rows; **reflection** – the shape inverts between columns (both shapes 'point' upwards in the first row and downwards in the second); **shading** – the shading at the base of the shape changes between columns.

53 c **number** – the number of shapes in the left-hand column equals the number of concentric shapes in the right-hand column; **shape** – the shape of the concentric elements matches the small shapes in the left-hand column; **shading** – the shading of the concentric lines matches the shading of the shapes on the left, i.e. if there are two white and two black circles, there must be two white and two black circles in the right-hand picture.

54 b **symmetry** – the grid works in a symmetrical pattern with the opposite corners containing similar elements; **shape** – the triangles match on opposite corners; **number** – the number of circles matches on opposite corners; **rotation** – the inner triangle doesn't change in the corners, but rotates around the grid so the shading will be on the left.
Distractors: **shading** – the shading of the circles matches on opposite corners although there

are no incorrect answer options; **general** – the elements in all the other squares are distractors.

55 e **shape** – (a) the large shape changes following the diagonals, looking from top left to bottom right, (b) the small shape changes between rows; **line style** – the line style of the large shape changes following the diagonal top right to bottom left; **shading/pattern** – the shading/pattern of the inner shape changes between columns.
Distractor: **shape** – the inner shape doesn't have a relationship with the outer shape.

56 a **position** – the position of the middle curled shape changes following the diagonals from top right to bottom left; **shape** – (a) the shape in the top-left corner changes between columns, (b) the shape in the top-right corner changes between rows.
Distractor: **rotation** – the small shapes do not rotate.

57 c **proportion** – the shapes enlarge, between columns, moving from left to right; **shape** – the large shapes change between rows; **shading/pattern** – the shading/pattern of the small circles changes between rows; **line style** – the line style of the outer shape changes following a pattern on the diagonal from top left to bottom right; **position** – the circles move one corner clockwise moving left to right between columns.
Distractor: **number** – the number of small circles changes between rows but they follow the same principles of movement and are not present in the answers.

58 e **shape** – the shape changes between 'columns'; **shading** – the shading changes direction in alternate cells moving around the edge of the picture.
Distractor: **shape** – the shape of the grid makes it difficult to follow the patterns.

59 a **number** – the number of concentric shapes increases by one, working clockwise around the outside of the picture; **rotation** – the inner shape rotates clockwise, as the shapes are added (this is clear if you follow the gap in the shape); **line style** – the new shape is always the outer shape each time, and this follows the order of line style shown in the cell in the top left, working down the lines shown; **proportion** – as a new shape is added, the other shapes reduce in proportion to allow the new shape to fit.

60 d **translation** – the top row of circles overlays on the middle row to create the pattern on the bottom row. When the circles overlap they become white, so only uniquely positioned black circles remain.
Distractors: **general** – there are many possible directions and options to choose, hence the difficulty of this question.

Paper 10 (page 18)

1 d **symmetry** – all the pictures are symmetrical apart from **d**.
In common/distractors: **shape** – the other features of the curved shape are random; **line style** – the line style of the square is random.

2 b position – in all the other pictures, the black circle is on the jointed line between the white circles.
In common/distractors: **number** – (a) there are always three lines; **angle** – the angles between the lines are random.

3 e angle – (a) the double arc marks the largest angle in the triangle, (b) the single arc marks the smallest angle, except for **a**.
In common/distractors: **angle** – some triangles include right angles, but this is random; **position** – the positioning of the triangles is random.

4 e rotation – when all the figures are rotated to the same position, for example with the large tinted triangle in the position shown in **a**, answer **e** is the only picture that is reflected vertically through the central line. Look at options **a** and **e** to see this relationship. To check the other pictures, rotate **b** 90° anticlockwise, **c** 180°, and **d** 90° clockwise.
In common/distractors: **shading** – the shading on the large triangles is random.

5 c number – the number of sides in each picture, when added together, is seven. The number of sides in **c** adds up to eight.
In common/distractors: **shading** – (a) one of the separate shapes is always shaded, (b) the shading on the circles is random; **shape** – there is no relationship between the two large shapes other than the number of sides; **number** – the number of small circles is random.

6 a position – the triangle 'points' into the circle except in **a** where it 'points' outside the circle.
In common/distractors: **shading** – the shading on the triangle is randomly inside or outside the circle; **position** – the position of the cross inside the small shape is random; **shape** – the small shape inside the circle is unimportant.

7 c direction – the arrow points to a black circle in all the boxes, apart from **c** where it points to a white circle.
In common/distractors: **shading** – (a) there are always two black circles and one white circle, (b) the shading on the arrow head is random; **position** – the corner the arrow comes from is unimportant.

8 d position – the triangle is always sat on one end of the thick line (left when the picture is rotated so that the line is horizontal and the shape is above the line) and the rectangle on the other, except for picture **d** where the rectangle is nearer to the centre.
In common/distractors: **shading** – the shading of the circle is random; **position** – the position of the circle is unimportant; **shape** – the outer shape doesn't follow a rule; **line style** – the line outside the main shape doesn't change.

9 d position – the shaded quarter of the small shape is always opposite the shaded triangle, except for picture **d** where it is on the same side.
In common/distractors: **position** – (a) the positions of the circle and square are unimportant, (b) the position of the shaded triangle (either top or bottom) is unimportant, (c) the position of the triangles to the left or right is not important; **line style** – the line outside the triangle is not important.

10 a shading – each shape has either a large triangle with vertical-striped shading or a rectangle with horizontally-striped shading, except for **a** where the rectangle has vertical stripes.
In common/distractors: **shading** – (a) the shading on the small triangles is random, (b) the shading on the circles is random; **number** – the number of circles is random; **position** – the position of the small triangle is unimportant; **direction** – the direction the large triangle is pointing is random.

11 d reflection – the shapes reflect in a horizontal line; **line style** – the line style of the two central shapes swaps over.

12 a translation – the lines assemble to create the large shape; **shading** – (a) the left-hand circle gives the shading at the bottom of the shape, (b) the middle circle gives the shading on the left-hand side of the shape and the right-hand circle the shading on the right-hand side of the shape.

13 c position – the two layers in the picture swap; **shading** – the shading on the two shapes in each figure swaps; **proportion/rotation** – the upper shape enlarges and rotates 45°.

14 b proportion/reflection – the three shapes change in size so that they can fit one inside the other and the whole picture reflects in a horizontal line; **shading** – (a) the bottom shape provides the shading for the largest shape, (b) the middle shape provides the shading for the smallest shape with the top shape then providing the shading for the medium-sized shape.

15 e position – the shapes move following the same rule in both pictures: (a) top right to top left, (b) top left to bottom left, (c) bottom left to top right, bottom right remains unchanged.

16 b rotation – the entire picture rotates 90° clockwise, so the large triangle will now be pointing to the left; **line style** – the style of the angled line changes from solid to dashed; **shading** – (a) the shading on the large triangle moves to the next triangle moving in a clockwise direction, so in the answer, the black hexagon will appear on the bottom triangle on the left, (b) the shading from the other two triangles also moves clockwise, so the answer must be **b**.

17 e reflection – the large shape reflects in a horizontal line; **proportion** – (a) the inner shape divides into two, (b) the small shapes all divide in half (with the bottom half inverting to match the top) and sit alongside the larger shape; **shading** – the shading from the top small shape goes to the right-hand side of the divided shape; **line style** – the line style of the outer and inner shapes swaps.

18 e reflection – (a) the top shape turns upside down, as if reflected in a horizontal line, (b) the bottom shape turns upside down as well and the two shapes join up; **shading** – the shading in the small shape at the top moves into the bottom shape.

19 c **rotation** – (a) the arrow rotates 90° anticlockwise, (b) the outer shape rotates 90° clockwise; **line style** – the line style of the arrow and outer shape swaps; **number** – the number of small circles in the arrow increases by one; **shading** – the small circles change from white to black.

20 c **rotation** – (a) the large outer shape rotates 90° clockwise, (b) the shape at the top end of this shape rotates 90° anticlockwise; **shading** – the shading of the small outer shape moves to the inner shape; **position** – the small shape moves to the open end of the large shape; **line style** – the line style of the large shape swaps with the inner shape.

21 e When reflected vertically, the white rectangle with the black cross moves from top left to top right, the white triangle moves to the left of the line and the black rectangle moves to the right at the bottom. Each shape changes to the opposite shade (so the top rectangle will have white cross on a black background).

22 b When reflected vertically, the white rectangle with the black cross moves from top left to top right, the white lozenge shape moves from bottom right to bottom left. Each shape changes to the opposite shade (so the top rectangle will have white cross on a black background and the lozenge will be white with a black circle).

23 c When reflected vertically, the two white triangles swap sides (the top triangle keeps the horizontal line at the top) and the black and white rectangle moves over to the bottom left. Each shape changes to the opposite shade (so the top triangles both become black and the rectangle becomes white with a black centre).

24 e When reflected vertically, the quadrilateral moves from top left to top right with the shortest side to the central line and the triangle moves from bottom right to bottom left. Each shape changes to the opposite shade (so the quadrilateral becomes white and the triangle is black with a white cross).

25 d When reflected vertically, the triangle moves from top right to top left, keeping the horizontal line at the base; the lozenge moves from the middle right to middle left and the rectangle moves bottom right to bottom left (the white triangle doesn't turn upside down). Each shape changes to the opposite shade (so the top triangle becomes black, the lozenge becomes white with black circle outlines and the rectangle becomes white with a black triangle).

26 a When reflected vertically, the lozenge moves from top right to top left, the rectangle moves from middle left to middle right and the triangle moves bottom right to bottom left (keeping the horizontal line at the base). Each shape changes to the opposite shade (so the lozenge becomes black with a white strip, the rectangle is white and the triangle is black).

27 d When reflected vertically, the triangle moves from top right to top left (keeping the horizontal line at the top), the quadrilateral moves middle left to middle right (keeping the short edge to the

central line) and the bottom triangle moves to the left (keeping the horizontal line at the base). Each shape changes to the opposite shade (so the top triangle becomes white, the quadrilateral becomes white and the bottom flag is black against the central line and white at its tip).

28 b When reflected vertically, the triangle moves to the top left (keeping the horizontal line at the top), the lozenge moves from middle left to middle right and the parallelogram moves from bottom left to bottom right (and continues to point upwards). Each shape changes to the opposite shade (so the triangle becomes black, the lozenge becomes white with a black circle and the parallelogram also becomes black).

29 c When reflected vertically, the top triangle moves to the top left (keeping the horizontal line at the base), the curved shape moves from the left to the right, the parallelogram from the left to the right (and continues to point downwards) and finally the triangle at the base points to the left. Each shape changes to the opposite shade (so the top triangle becomes black along the central line and white at the tip, the curved shape becomes black with a white circle, the parallelogram becomes white and the bottom triangle black).

30 d When reflected vertically, the top triangle moves to the top left (keeping the horizontal line at the top), the central triangle moves middle left to middle right and the parallelogram moves from bottom left to bottom right (and continues to point upwards). Each shape changes to the opposite shade (so the top triangle becomes black, the central triangle becomes white with a black cross and the parallelogram becomes black along the central line and white at the tip).

31 e The most identifiable feature of the small quadrilateral is the angle on the right. Options **b**, **c** and **e** have a similar angle although neither **b** nor **c** has the small vertical line on the right-hand side. This leaves option **e**, which is the correct answer.

32 a The most identifiable feature of the quadrilateral is angle of the side on the left (since the bottom two sides appear to match up with the bottom of all the hexagons). Options **a** and **e** have a similar angle although the top side in **e** is too steep, as it forms one side of the hexagon. This leaves option **a**, which is the correct answer.

33 c The most identifiable feature of the five-sided polygon is the angle of the side on the right.

Options **b** and **c** have a similar angle although there are no lines within option **b** matching the short angled side on the left. This leaves option **c**, which is the correct answer.

34 **e** The most identifiable features of the small quadrilateral are the symmetrical angled sides to the left and right. Options **a** and **e** both have similar angles on the right, although there are no lines within option **a** matching the angled side on the left. This leaves option **e**, which is the correct answer.

35 **a** The most identifiable feature of the polygon is the stepped shape at the centre of the base. Options **a**, **c** and **e** all have vertical lines at the centre of the shape, although in **a**, the step is to the right of centre and in **c** the step isn't present. This leaves option **a**, which is the correct answer.

36 **d** The most identifiable feature of the six-sided polygon is the angle formed between the top and right-hand sides. All the options aside from **e** have a similar angle although only **c** and **d** have a right angle to the left, similar to that in the polygon. Looking closely at **c**, the proportions of the square shape on the left are too large, leaving option **d**, which is the correct answer.

37 **b** The most identifiable feature of the polygon is the curved edge at the top, which gives a clue to where it will appear in the picture below – towards the right, so that it follows the outline of the larger shape. Option **b** is the only possible answer as this is the only picture with two vertical lines that span this part of the shape, so this must be the correct answer.

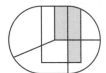

38 **c** The most identifiable feature of the six-sided polygon is the angle on the left-hand side of the shape. Options **a**, **b**, **c** and **e** have a similar angle,

although only **b**, **c** and **e** have a horizontal rule above it. As **b** and **e** don't have the rectangular shape underneath, this leaves option **c**, which is the correct answer.

39 **d** The most identifiable feature of the five-sided polygon is the 'prong' shape at the base. Options **a**, **b**, **d** and **e** have a similar shape, so you need to look more closely at each one. In **a**, the prongs are not of even lengths, **b** doesn't have a vertical line to the left, and **e** doesn't have a diagonal line on the left. This leaves option **d**, which is the correct answer.

40 **b** The most identifiable features of the six-sided polygon are the angle and step at the top of the shape. All of the options have similar features, although **a** doesn't have the correct length side on the right and the base; the bottom left-hand side is missing in **c**; in **d** none of the possible shapes is long enough along the vertical side and in **e** the right-hand side of the shape cannot be formed with any combination of lines. This leaves option **b**, which is the correct answer.

41 **d** **number** – (a) the code in the upper small box represents the number of circles: F is three, G is two, (b) the code in the lower small box represents the number of black balls: X is one, Y is two.

42 **a** **shading** – the code in the upper small box represents the shading of the large shape: R is diagonal stripes from top right to bottom left, S is horizontal stripes and T is vertical stripes; **shape** – the code in the lower small box represents the large shape: L is a diamond and M an oval. Distractor: **position** – the position of the small shapes is unimportant.

43 **e** **direction** – the code in the upper small box represents direction of the angle at the bottom: X pointing upwards, Y is pointing downwards and Z is pointing to the left, (b) the code in the lower small box represents the direction of the loop at the top: P is for the open side pointing to the right, Q pointing to the left. Distractor: **position** – the position of the cross within the circle is unimportant.

44 **a** **shape** – the code in the upper small box represents the large shape: K is the bullet shape,

L the six-sided polygon and M the diamond; **line style** – the code in the lower small box represents the vertical line style: P is for the wavy line and Q for the straight line.

45 b **position** – the code in the upper small box represents the position of the rectangle: R is on the left of the box, S is in the middle and T is on the right; **line style** – the code in the lower small box represents the line style of the rectangle: L is for dashed, M is for solid and thick, N is for solid and thin.
Distractors: **position** – the triangle and its position are unimportant; **shading** – the shading of the rectangles is unimportant.

46 c **direction** – the code in the upper small box represents the direction the large triangle is pointing: F is down, G is up, H is to the right – as there is no code for pointing to the left, this must be J; **shape** – the code in the lower small box represents the small shape: W is for a triangle, X a circle, Y a rectangle and Z a square.
Distractors: **shading** – the shading of the triangle is unimportant; **line style** – the style of the single line is unimportant.

47 b **shape** – the code in the upper small box represents the central shape: F is for the circle, G the hexagon, and H the square; **line style** – the code in the lower small box represents the line style of the rectangle: P is for dashed, Q is solid and thick, R solid and thick and S dotted.
Distractors: **number** – the number of dots is unimportant; **shading** – the shading of the top corner is unimportant.

48 b **line style** – the code in the upper small box represents the line on the left: L is wavy, M zig-zag and N square; **position** – the code in the lower small box represents the position of the square: X is at the top, Y is in the middle and Z at the base.
Distractor: **shading** – the shading in the square is unimportant.

49 d **shape** – the code in the upper small box represents the bold line or shape: W is a triangle, X a 'U' shape and Y an 'R' shape; **position** – the code in the lower small box represents the position of the triangle with the dashed sides: in Q the long side next to the right angle is horizontal and at the top, in R this side is vertical and to the right with the right angle at the top, in S it is also vertical and to the right but with the right angle at the bottom, in T it is horizontal with the right angle to the left above this side.
Distractors: **shape** – neither of these elements is part of the code; (a) the small shapes, (b) the second triangle; **shading** – the shading on the triangle is also not relevant.

50 e **number** – the code in the upper small box represents the number of small horizontal lines on the left: R is three, S is two, T is four; **position** – the code in the lower small box represents the position of the triangles in relation to each other: in L they are pointing towards each other, in M

they are both pointing upwards, in N they are pointing away from each other.
Distractor: **line style** – the thickness or style of the vertical line is unimportant; **shading** – the shading on the triangles is unimportant

51 e **shape** – this is an alternating sequence between a cross and a triangle with a circle inside; **shading** – the shading repeats in every alternate shape, so as the first cross is white on the outside, then black on the inside; the third cross will also follow this shading.
Distractors: **shading** – the sequence could potentially follow a longer sequence of tints although the sequence given here shows two alternating patterns already which should be spotted.

52 d **rotation** – the square rotates 135° clockwise so if the vertices are pointing to the sides, on the next rotation they will be pointing to the corners of the box; **position** – the white circle moves one corner clockwise around the box, working left to right across the sequence.

53 b **proportion** – (a) the small hexagon in the box on the left increases in size as it moves across the sequence, (b) in each box, one sixth of the shape disappears losing a side in a clockwise direction, (c) an additional hexagon then appears within the growing hexagon with one side being added in an anticlockwise direction, beginning in box 2 (this hexagon doesn't increase in size); **rotation/position** – the small triangle rotates 90° clockwise as it moves around the box, one corner clockwise each time.
Distractor: **shading** – the triangle alternates between black and white although there are no incorrect answer options shown.

54 d **number** – (a) the number of lines decreases by one moving left to right across the sequence, (b) the number of circles increases by one with an extra circle added to the bottom end of the set; **shading** – the shading on the circles alternates between black and white from one box to the next (including the circle broken by the line).
Distractor: **shading** – the extra circle appears to be added beneath the split circle, although looking at the final box shows that the shading swaps.

55 e **rotation** – the arrow rotates 135° clockwise, moving left to right across the sequence; **shading** – (a) the shading on the arrow head alternates between black and white, (b) the shading inside the arc alternates between white and black.
Distractor: **position** – the position of the arc moves 45° anticlockwise across the sequence although no answer options place this incorrectly.

56 a **reflection** – the picture reflects in a vertical line as it moves across the sequence (so the pattern alternates); **shading** – (a) the black circle moves down the shape by one 'point' each time, (b) the circle with the cross moves up the sequence two 'points', and when it reaches the top, then goes down again (so in the last box, it is covered by the

black circle), (c) the white circle with the black centre alternates between the bottom two points, (d) the white circles fill the remaining points.

57　c　**rotation** – this is a progressive sequence where the rotation of the triangle increases by 45° as it moves clockwise across the sequence, so the first turn is 45°, then 90° then 135° so the final rotation will be 180°; **position** – (a) the square moves around the box by one corner anticlockwise, (b) the cross in the square moves between the corners and the sides in alternate boxes, (c) the shading in the circle swaps between being inside and outside the triangle.

58　c　**rotation** – the arrow rotates backwards and forwards between the top right-hand and bottom left-hand corners (so all the arrows in the bottom row point up to the right); **position** – the shading in the circle moves between the top and bottom quarters (which is clear in the sequence along the top row), (b) – the shading on the arrow alternates between top and bottom halves.

59　e　**shape** – (a) the shape alternates between a circle and square, so all the shapes on the top row are circles, (b) the small white shape in the corners changes every two boxes, so the pattern is two squares, two quarter circles, two rectangles; **position** – (a) the small circle moves one corner anticlockwise working left to right across the sequence, (b) the shaded quarter of the shape moves one quarter clockwise moving left to right across the sequence.
Distractor: **position** – the movement of the white shape in the corner doesn't follow a clear rule, apart from changing position every second box, however there are no incorrect answers.

60　b　**proportion** – the shaded 'L' shape reduces in size, moving left to right across the sequence; **number** – one extra arrow is added, moving left to right across the sequence; **position** – the shading in the 'L' alternates between the bottom the two 'arms' although the shading doesn't change direction; **direction** – (a) the arrow at the base alternates between pointing to the left and right, (b) the arrow within the white 'arm' always points to the edge of the box, (c) the arrow outside the shaded 'arm' always points to the corner of the 'L' shape, (d) when the extra arrow is added, it always points in the opposite direction to the previous one in the previous box.

Paper 11 (page 27)

1　c　**shape** – each large shape has a smaller, identical shape inside it; **line style** – the line style of the outer shape is a short dash
Distractor: **shape** – the number of sides and other properties of the shapes are unimportant.

2　a　**shape** – shape is an isosceles triangle with half of the base missing and circles in each point; **shading** – one circle is shaded black, one white and the other partly shaded black; **position** – (a) the white circle is always on the line ending in the

middle of the 'base' of the triangle, (b) the shading on the circle sitting on the angle is always inside the triangle.
Distractor: **reflection** – whether the open side is on the left or right (when the apex is pointing upwards) is not important.

3　b　**number** – each shape has five sides; **symmetry** – each shape has at least one line of symmetry.
Distractors: **shading** – the side with the shaded band is unimportant, as long as it doesn't break the symmetry; **line style** – the line style of the shape is unimportant.

4　d　**number** – the number of shapes in each picture equals the number of sides in the shapes within it, i.e. if the small shapes have four sides, then there will be four identical shapes in the picture; **position** – (a) the lines joining the small shapes meet in the centre, (b) the two crosses in each are positioned differently with one running diagonally, the other vertically and horizontally; **shape** – the sides of all the shapes are of equal length.

5　e　**shape** – the shapes are all completely curved; **number** – there are always two circles, one triangle and a diamond; **shading** – (a) one of the circles is always black, (b) one other shape is also shaded black, either the circle or triangle, but never the diamond.
Distractors: **position** – the position of the small shapes in relation to each other is not important; **shape** – no other features of the shapes are important (such as symmetry or position).

6　c　Once folded into a cube, the triangles will be pointing towards each other. When **c** is folded into a cube, the four central faces work as a loop and join up, so the triangles will be pointing towards each other with the square face to the left. Turning the net through 90° clockwise can make this easier to see. Notice net **d** will not make a cube.

7　d　Once folded into a cube, the points of the stars will be next to each other. When **d** is folded into a cube, the four central faces that touch continue to touch so these sides are always next to each other, the circle then folds down to touch both faces. Turning the net through 90° anticlockwise can make this easier to see.

8　e　Once folded into a cube, the black triangle will point to the side of the white triangle and the circle sits at the base of the white triangle. When **e** is folded into a cube, the triangles on the central line will continue to be next to each other, the circle then folds down to touch both faces. Turning the net through 180° can make this easier to see.

9　d　Once folded into a cube, the black triangle will point away from the cross. When **d** is folded into a cube, the triangle and the star will continue to be next to each other, and stay in the same direction as they are on the central line. When the net is rolled into a cube the cross folds inwards to form the top. Turning the net through 90° anticlockwise can make this easier to see, **e** is incorrect because the triangle will point towards the cross when the net is folded.

10 b Once folded into a cube, the white triangle will be pointing towards the black star. When **b** is folded into a cube, the black star and circle will be next to each other on the central line. This means that when the triangle folds inwards down, it will point to the face next to the circle, which will be the star. Turning the net through 90° anticlockwise can make this easier to see.

11 a The fold lines act like diagonal lines of reflection, so when unfolding the second fold in the example, there will be a hole in a diagonal beneath and to the right the same distance above the diagonal line as the visible hole is above it. Undoing the first fold will give a reflection of these two holes above the long diagonal line so that a diamond pattern appears.

12 d The fold lines act like lines of reflection, so when unfolding the second fold in the example (which works like a horizontal reflection), there will be a hole directly above the dashed line the same distance away that the visible hole is beneath it. Undoing the first fold will give a reflection along the diagonal and so there will be two more holes on the top edge of the square.

13 c The fold lines act like lines of reflection, so when unfolding the third fold in the example (which works like a vertical reflection), there will be two holes to the right of the dotted line, the same distance away as the visible holes are on the left. Undoing the second fold works like a diagonal reflection so the single hole (see diagram on the right on page 31) will reflect and appear in the bottom left corner. Undoing the final fold will reflect all six holes in the horizontal fold line so there will be 12 holes following the pattern shown in **c**.

14 e The fold lines act like lines of reflection, so when unfolding the third fold in the example (which works like a diagonal reflection), a second circle will appear to the right and above the dashed line the same distance away as the visible hole is beneath it. Look carefully at the second fold – this only creates a double layer half way down the rectangle so only the top circle and the heart-shaped hole are reflected in the horizontal line (so the heart shape will appear upside down) above the central line of the square. Undoing the final fold will reflect these last two holes so the heart appears the right way up again on the left-hand side opposite the circle on the right, as shown in answer **e**.

15 c Try counting the holes created as you 'unfold' this example to help you check the answer. The fold lines act like lines of reflection, so when unfolding the third fold (which works like a vertical reflection), a circle will appear on the right-hand side of the dashed line the same distance away as the visible circle is on the left. Undoing the second fold creates a diagonal reflection in the dashed line, so look carefully at the vertices on the small triangle – the vertex close to the dashed line will be the same distance away beneath the line

(so the triangles will appear to point towards each other); the two circles in the centre will reflect at 90°, so be in a vertical line from the base; the reflection of the top circle will appear towards the bottom right-hand corner of the large triangle. Undoing the final fold will reflect all of the holes in the dashed line on the first diagram creating the symmetrical pattern shown in **c**.

16 d **line style** – the first letter represents line style: R is dashed, S is solid and thick, T is solid and thin; **shape** – the second letter represents shape: L is oval, M a triangle, N a rectangle and O a diamond.

17 b **direction** – the first letter represents the direction the triangle is pointing: P is up, Q is down; **shape** – (a) the second letter represents the shape of the line: X is curved, Y is square, Z is zig-zag, (b) the third letter represents the left-hand shape: L is a circle, M a square.
Distractor: **position** – the position of the line (either down then up or vice versa) follows the rule for the first letter but there are no answer options that work for this solution and so it is a distractor.

18 a **shape** – the first letter represents the middle shape: P is a triangle, X a bullet shape, Y a rectangle; **shading** – (a) the second letter represents the shading on the middle shape: X is horizontal stripes, Y is diagonal stripes, (b) the third letter represents the shading on the left-hand shape: F is white, G is black.
Distractor: **shading** – the right-hand shape and its shading are not related to the code.

19 b **direction** – the first letter represents the 'direction' the large open shape is pointing in: P the opening is to the right, Q is up, R is left, S is down; **line style** – the second letter represents the line style of this shape: L is solid and thick, M is solid and thin, N is dashed; **number** – the third letter represents the number of vertical lines: X is three, Y is two, Z is four.
Distractor: **shape** – the small shapes are not related to the code.

20 e **shape** – the first letter represents the shape created where the two larger shapes cross over: F is a diamond, G is a rugby-ball shape, H a smaller diamond; **line style** – the second letter represents the line style of the left hand shape: L is solid, M dashed; **number** – the third letter represents the number of small black squares: W is two, X is three, Y is one.
Distractors: **line style** – the line style of the right-hand shape is random; **shape** – (a) the larger shapes are unimportant apart from the shape they create where they cross over, (b) the small left-hand shape is unimportant; **shading/pattern** – the pattern on the left-hand shape is unimportant.

21 d The small shape is the same size as the piece that will be removed and it is also the same way around. The angle of the 'cut' in shape **d** is the same as the angle of the sides in the small shape that has been removed.

22 c The small shape is the same size as the piece that will be removed and it is also the same way around. The angle of the 'cut' in shape **c** is the same as the angle of the sides in the small shape that has been removed.

23 e The small shape is the same size as the piece that will be removed and it is also the same way around, fitting in the angle to the right of shape **e**.

24 e The small shape is the same size as the piece that will be removed and it is also the same way around. It fits into the centre of **e**. This answer is correct because the angle at the top left of the shape completes the arrow correctly and the horizontal line at the centre of the shape is the correct length to fit the missing piece.

25 d The small shape is the same size as the piece that will be removed and it is also the same way around. It fits into the base of **d**. This answer is correct because the cut-out section at the bottom of this shape matches the top of the missing piece and the triangle on the right is the correct size to match that in the complete shape on the left.

26 c **shape** – the shapes change between rows; **position** – (a) the position of the shaded shape changes between columns (left/top, middle, right/bottom), (b) the position of the line changes following the diagonal top left to bottom right; **line style** – the style of this line changes following the diagonal top right to bottom left.

27 d **proportion** – the circle enlarges as it moves clockwise around the boxes in the hexagon; **position** – the cross alternates between a diagonal and vertical/horizontal position; **reflection** – shading in opposite circles reflects and goes to the opposite corner.
Distractors: **rotation** – the cross appears to rotate, but following the tint makes it clear that this cannot be the rule; **shape** – the other elements in the grids are not relevant.

28 e **rotation** – the pattern works in columns: the thick black arrow rotates 45° clockwise as it moves down the column; **shape** – the small shape works in a diagonal pattern from top right to bottom left.
Distractors: **rotation** – the other two arrows rotate differently, (a) the plain arrow rotates 90° clockwise as it moves down the column, (b) the double-headed arrow rotates 45° clockwise in the left-hand column and 90° clockwise in the right-hand column.

29 a **number** – the number of circles increases, moving clockwise around the hexagon; **position** – the circles always begin by the side of the base of the triangle shape and work in towards the centre; **line style** – the line alternates between solid and dashed; **shape** – the larger shape alternates between a triangle and a hexagon; **shading** – the shading works in pairs, so the missing triangle will take on the shading shown in the bottom section.

30 c **shading** – the shading on the large figure and central bar alternate around the outside of the shape; **position** – the central bar takes its position from the cell diagonally opposite to it; **rotation** – the two arrows rotate clockwise around the hexagonal cell moving two vertices around the sides.
Distractors: **rotation** – (a) at first, the arrows appear to follow a rotation rule linked to the inner shape, which they do not, (b) because the arrows rotate, the central bar can also be assumed to rotate, although it actually follows the rule of position described above.

Paper 12 (page 35)

1 c The bottom layer of the stack has four cubes at the front, then a gap and two cubes at the back; the first two cubes at the back line up with the first two cubes at the front. When looked at in plan view this will show four squares at the bottom left of the plan, then two squares at the top left. The second layer at the front doesn't alter the plan, but the stack behind fits between the two rows and is one cube in from the right. This must fit on the middle row of the plan, on the second square from the right.

2 a The bottom layer of the stack works in an 'L' shape, with another cube in a separate stack. As the 'L' is three cubes deep these will begin the three rows on the plan with another two squares along the top. Looking at the other layers, there are no cubes that hang outside this shape. The separate stack is not in line with the final cube at the back of the 'L' so must be further back. However, it does line up with the front of the 'L' so will be represented in the bottom right-hand corner of the plan.

3 d The bottom layer of the stack has three cubes on the front row (two, then a gap then one), so the bottom row of the plan will have two squares on the left and one on the right. There are two cubes running backwards from the second cube and then another one to the right at the back; this upside down 'L' shape will mean all three squares on the second column of the plan are filled in and the top square on the third column. The final square on the bottom layer runs back from the right-hand cube so will be the last square on the middle row of the plan. There are two cubes on the top layer that do not line up with the bottom layer: the left-hand cube (which will fit in the top-left corner of the plan) and the cube in the second row, which will fit in the third square of the second row.

4 b The bottom layer of the stack has four cubes at the front, filling up all four squares in the bottom row of the plan. You can therefore ignore the stack on the front row, as this can't add to the plan. There is then one cube running back from the second cube on the left (the second square on the middle row of the plan). As the final square on the bottom layer is only attached by a side, and space can be seen to the right and above it, this cube must be in line with the front cube on the left (and therefore the top left-hand square of the plan). However, the two extra cubes on the top layer overhang these gaps filling up the left-hand

square in the middle and the second square on the top row of the plan.

5 c The bottom layer of the stack has three cubes on the front row (one, then a gap then two), so the bottom row of the plan will have one square on the left and two on the right. There are then two cubes running back from the first cube on the left (filling the first column of the plan). There is then a further cube that can just be seen coming from the central cube, which will be the centre square on the second column. The stack on the left shows three cubes on the top layer, two that cover the cubes on the base, but the top of the 'L' shape also covers the top square on the second column of the plan. Finally the stack at the back has two cubes, one that hangs over into the second row and so will add an extra square to the end of this row to complete the plan.

6 a The 'L'-shaped block has been rotated anticlockwise through 90° clockwise and then rolled backwards to sit on its long edge. Two single cubes are then stacked beside it at the back, one on top of the other. The two-cube cuboid then sits across the top of the 'L' and the single cube. Finally the three-cube cuboid stands on its end at the left and the final single cube sits in a corner of the 'L'.

7 e The left-hand 'L'-shaped block doesn't change position, but the right-hand 'L' shape rotates 90° forwards and sits over the corner of the front 'L'. The two-cube cuboid then sits at the back behind the front 'L' with the two single cubes placed at the left-hand side and on top of the stack at the back.

8 e The two-cube cuboid rotates 90° so that the end is facing forward. The 'L'-shaped block then sits on top of this so that only the front section of the cuboid is showing. The 'T'-shaped block then rotates 90° around a vertical line and rolls 90° backwards so the top of the 'T' is facing forward; this slides next to the 'L' on the left to form the top of the shape and one of the single cubes sits beneath it at the back so that it doesn't fall over! Finally, the remaining single cube sits at the front.

9 d The three-cube cuboid remains in the same position and the two 'L' shaped blocks hang over the back of it (so have both rolled backwards 90°). The two-cube cuboid and single cube then stand at the front.

10 a The corner shape at the front of the stack is made up of four cubes, shown at the top left of picture a. This shape rotates 180° around a vertical line and then rolls towards the back so that it stands on one 'arm'. One of the 'L' shapes drops onto its side and then fits around the base of the four-cube block; a single cube is then placed behind it at the back right of the picture. The second 'L' shape rolls towards the back and sits on top of this single cube and over the top of the four-cube block. Finally the remaining single cube and the two-cube cuboid sit on top to complete the stack.

Paper 13 (page 37)

1 c **number** – each figure has an odd number of sides; **shape** – one internal angle is marked with a curved line.
Distractor: **symmetry** – the shapes are symmetrical but the angle does not necessarily follow this symmetrical line.

2 b **shading** – the shading on the triangle always runs at a right angle to the side of the square it is touching; **position** – (a) the circle is always to the right of the triangle, when the triangle is pointing upwards, (b) the arrow always comes from the corner to the left of the triangle when it is pointing upwards.
Distractor: **direction** – the direction of the arrowhead is unimportant.

3 e **symmetry** – the outline of the shield is always symmetrical; **position** – the top-left portion of the shape is always shaded black.
Distractors: **number** – the number of curves/points at the top of the shield are not important; **shape** – (a) the shape of the bottom of the shield is not relevant, (b) the shape of the central pattern isn't important, **proportion** – the proportions of the four sections of the inner shape are not important.

4 a **size** – there is a large and small circle in each figure; **line style** – (a) each figure has an outer dashed rule and an inner solid rule, (b) the large circle has a bold outline, (c) the small circle has a standard solid outline; **position** – the circles need to be separate from one another.
Distractor: **position** – the position of the circles, within the shape, in relation to each other is unimportant.

5 d **shape** – there is a circle, diamond and triangle in each figure; **position** – (a) the diamond is always the central shape, (b) the top shape always has a thick line style.
Distractors: **position** – (a) the position of the triangle and circle in relationship to each other is random; **proportion** – the length of the lines is random.

6 a **translation/proportion** – the shapes all move to fit together: (a) the left-hand shape is the largest, (b) the bottom shape is medium-sized and therefore the right-hand shape is the smallest.
Distractor: **shading** – the shapes keep their original shading although the answer options provide many variations.

7 c **rotation** – arrow shape rotates 90° clockwise, so the arrow in the second picture will be pointing to the left; **proportion** – the shape at the top on the right enlarges and fits around the arrow; **shading** – the shading of the small shape at the bottom moves to the arrow; **position** – the two small shapes on the bottom right move to sit on either side of the arrow.

8 c **line style** – (a) the line in the centre gives its line style to the shape above it, (b) this top shape then gives its style to the bottom shape; **proportion** – this bottom top shape then enlarges to fit around

the other shapes; **position** – the shape on the left moves to the centre but otherwise remains unchanged.

9 **b** **rotation** – both shapes rotate 90° with the right-hand shape moving above the left; **line style** – the two line styles within each shape swap (so inner becomes outer and vice versa); shading – the shading swaps between the shapes.

10 **e** **proportion** – (a) the large shape reduces in size to become the shape on the bottom left, (b) the top-left shape becomes the larger shape on the right; **reflection** – this large shape reflects in a vertical line so that it is 'pointing' in the opposite direction; **number** – the number of wavy lines in the large shape equals the number of sides in the shape on the top left (so four wavy lines equals four lines, so the shape is a square); **shading** – the shading of the shape on the bottom left moves to the shape on the top left.
Distractor: **rotation** – the shading doesn't rotate.

11 **c** When the net is folded to make a cube, if the face with the dashed square is facing forward, the face with the double squares will fold down to make the top and the face with the white circle will then fold in to make the left-hand face. In all the other options, the position of the faces in relation to each other is incorrect.

12 **d** When the net is folded to make a 3D shape, the triangular face with the white triangle pointing upwards will be on the left of the triangular face with the black triangle. This will mean that the square face attached to these faces will also be next to each other in the final shape.

13 **e** When the net is folded to make a cube (rolling it up along the central line), if the face with the plain square is facing forward, the face with the circle will be next to it and underneath. If this cube is then rolled over to the right, the circle will be on the left and the two black triangles will fold on top to make the upper face.

14 **a** When the net is folded to make a cube (rolling it up along the central line), if the face with the 'I' shape is facing forward, the face above it will be the 'U' shape. The 'U' will have its open side to the right since the net has not been rotated. The striped face will then fold upwards to form the side, with the stripes parallel to the tall bar of the 'I'.

15 **b** When the net is folded to make a cube, if the face with the black band down the side is facing forward, the face with the diamond will be folded back, out of sight. The face with the black triangle will then fold down to make the top and join with the cross, which rolls up from the other end of the net.

16 **a** The fold lines act as lines of reflection, with a separate line of reflection for each small shape, so: the triangle at the top left reflects horizontally, and points downwards with the right angle top left; the triangle on the left reflects vertically so will flip left to right; the triangle at the bottom

has equal angles so when it reflects horizontally will point upwards. Finally, the rectangle on the right will reflect vertically so just appear to move to the left.

17 **d** The fold lines act as lines of reflection, with a separate line of reflection for each small shape, so: the triangle at the top left reflects horizontally, and points downwards with the right angle top right; the triangle on the left reflects vertically so will flip left to right with the right angle now appearing on the left of the triangle; the square on the right will reflect vertically and touch the right-hand side of the top triangle.

18 **e** The fold lines act as lines of reflection, with a separate line of reflection for each small shape. The curved shape at the top reflects in a horizontal line so the shape appears upside down; the triangle on the left reflects vertically so appears to flip from side to side. This gives a horizontal line that joins to the bottom of the curved shape with a diagonal line in from the top-left corner. When the quadrilateral at the base is folded along the horizontal line, the top edge meets the base of the triangle above and continues the line across the whole square. Finally, when the two triangles fold in vertically from the right, they flip over making the same 'V' shape in reverse, as is shown in answer **e**.

19 **c** The fold lines act as lines of reflection, with a separate line of reflection for each small shape within the square, so: the triangle on the left will reflect vertically and appear to be pointing to the left with the horizontal at the base (the answer can now only be **b** or **c**); the triangle at the base reflects horizontally, so appears beneath the line with the right angle remaining on the left. Option **c** is the only answer with both triangles in the correct position so the square must be reflected horizontally.

20 **d** The fold lines act as lines of reflection, with a separate line of reflection for each small shape within the square. Looking at the answers, the quadrilateral on the right (which will reflect vertically) will have a horizontal base and a diagonal going from top left to bottom right before it is reflected, so answer **c** is incorrect. The triangle above must have its vertical side on the left, otherwise there would be an extra diagonal in the answer, so only **b** or **d** could be correct. The triangle at the base must also have its vertical side to the left, so the only possible answer is **d**.

21 **b** 90° is a quarter turn clockwise, which is shown in option **b**, **d** is rotated 90° anticlockwise.

22 **d** 45° is half of 90°, so the shapes will appear at an angle. As the square is above the vertex of the triangle, it must remain in this position (shown in option **d**). Option **b** shows the square beneath the triangle so is the wrong answer.

23 **a** 135° is 45° added to 90°, so any lines that are vertical will appear to point down to the right (see the central line in the picture on page 41).

This means that only options **a** and **d** are possible. As the small circle will move as the picture rotates, this will end up on the right-hand side so option **a** must be correct.

24 **e** 225° is 180° added to 45° so the whole shape will turn upside down and then angle up to the left. Looking at the square section on the left-hand side of the shape, this will mean the line currently top left will be the base of the new shape, making **e** the correct answer.

25 **b** 135° is 45° added to 90°, so imagine rotating the picture in two stages. Drop the 'Z' shape down on its side (rotating it clockwise) to imagine a 90° rotation; the top and bottom horizontal lines will now be vertical and the small right angle will be at the bottom left. Then imagine rotating it another 45°; the diagonal will now appear vertical with the right angle remaining on the left as shown in option **b**.

26 **a and d** **shape** – the pictures are made up of an isosceles triangle and a quadrilateral; **position** – the isosceles triangle meets the quadrilateral at its apex; the quadrilateral makes a right angle with the triangle on the short side beneath the angled side.

27 **b and c** **shape** – most of the pictures are made up of a right-angled triangle and a square with an open side, plus an angled line; **position** – the angled line is only matching in two shapes, **b** and **c** where it comes up from the corner at the base of the longest edge on the picture.

28 **c and e** **shape** – the pictures are made up of four triangles and a quadrilateral with a smaller triangle and circle; **position** – the larger triangle points to the right angle joining the smaller triangles in **a**, **c** and **e**. However, the small shapes swap position in **a** so the answers must be **c** and **e**.

29 **a and e** **shape** – the pictures are made up of a triangle with five jointed lines, two lines have small circles at the ends; **position** – the small circles are on the 'arms' coming out of the triangle base (as if they are beginning to form a hexagon) in **a**, **c** and **e**. However, in **c** the jointed 'arm' is on the left and in both **a** and **e** it is on the right, when the triangle base is at the bottom of the picture.

30 **b and d** **shape** – the pictures are made up of a triangle with five angled lines plus a small square and lozenge; **position** – the lozenge is only on the inner edge of the line in **c** so this cannot be an option, **a** is the only option where the lines coming out of the triangle vertex point in the same direction, so this cannot be correct either. In option **e** the small square and lozenge are on the same angled line, unlike in **b** and **d** so these two options are the only possible matches.

31 **b** **position** – (a) the circle is on the bottom-left side of the square (so the answer cannot be **a** or **c**), (b) the rectangle is placed over the right-hand vertex of the square (so the answer cannot be **d** or **e**).

32 **d** **position** – (a) the square on the left forms an obtuse angle with the triangle (so the answer cannot be **a**), (b) the circle forms a curved line from the bottom of the square and approximately half way down the triangle on the right (so the answer cannot be **b**, **c** or **e**).

33 **e** **position** – (a) curved shape forms two right-angles the same distance from the top and bottom of the rectangle (so the answer cannot be **b** or **d**), (b) this shape also stands out from the left-hand side of the rectangle (so the answer cannot be **c**), (c) the top apex of the trapezium is above the top of the rectangle (so the answer cannot be **a**).

34 **c** **position** – (a) the narrow rectangle sticks out of the left side of the triangle to a point in line with the centre of the circle on the right (so the answer cannot be **b**), (b) the circle doesn't change size (so the answer can't be **a**), (c) the quadrilateral sticks out of the bottom of the triangle (so the answer cannot be **d**), (d) the circle is near to the top of the triangle (so the answer cannot be **e**).

35 **d** **position** – (a) the triangle sticks out from the circle at 9 o'clock and 12 o'clock, if the circle is seen as a clock face (so the answer cannot be **a** or **c**), (b) the narrow rectangle is placed just to right of centre on the circle, (so the answer cannot be **b**), (c) the straight sides of the lozenge are clearly visible outside the circle (so **e** cannot be correct).

36 **a and c** Option **a** appears at the bottom right of the image, reflected in a horizontal line; **c** appears to the left rotated 45° clockwise.

37 **b and e** Option **b** appears at the top right of the image, rotated 135° anticlockwise; **e** appears to the left of centre, rotated 45°.

38 **b** **line style** – the first letter represents the line style: F is solid and thin, G dashed, H solid and thick; **shape** – the second letter represents shape: X is a curved 'U' shape, Y a 'V' shape and Z a square 'U' shape.

39 **c** **proportion** – the first letter represents whether there is a quarter of the shape cut out: L stands for cut-out, M for no cut-out; **shape** – the second letter represents shape: R a circle, S a square, T a hexagon; **position** – the third letter stands for where the small circle is placed: X is inside the large shape, Y is outside the large shape. Distractor: **line style** – the dashed line is unimportant.

40 **e** **direction** – the first letter represents the direction of the triangle in the centre of the shape: K stands for pointing upwards, L for pointing downwards; **line style** – the second letter represents the line style of the outer shape: X is solid, Y is dashed; **position** – the third letter stands for whether the circle is positioned on a corner or a side: F is on a corner, G is on the side.

Distractors: **shading** – the shading of the shapes is not important; **shape** – both the large shape and the smaller shapes at the side are unimportant.

41 **d** **line style** – the first letter represents the style of the inner line: L is dashed, M is solid; **position** – the second letter represents the position of the double angle shape: F is for the horizontal line at the top and pointing right, G is for the horizontal at the top and pointing left, H is for the horizontal at the bottom and pointing right, J is for the horizontal at the bottom and pointing left; **shading** – the third letter stands for the shading of the narrow rectangle: R is striped, S is black, T is white.
Distractor: **number** – the number of circles is unimportant; **position** – the position of these small circles is unimportant.

42 **a** **shape** – (a) the first letter represents the central section of the outer shape: F is circular, G is diamond-shaped, H is rectangular, (b) the second letter represents the outer portion of this shape: X is rounded, Y is pointed, Z has square ends, (c) the third letter stands for the inner shape: P is a circle, Q a diamond, R a hexagon and S a triangle.

43 **d** **line style** – the first letter represents the style of the connecting horizontal line: P is solid and thin, Q is dashed, R solid and thick; **shape** – the second letter represents the smaller shape on the right: W is a hexagon, X a circle, Y a diamond and Z a square (V is a triangle and not shown but a logical connection in the answer); **size/shading** – the third letter represents the style of circle on the left-hand side: L is a double circle, M a small and black, N larger and white.
Distractors: **shape** – the larger shape on the right-hand side is unimportant; **line style** – the line style of the outer circle on the left-hand side is unimportant.

44 **b** **proportion** – the circle adds one quarter in a clockwise direction, moving left to right across the sequence; **position** – the small circle moves from bottom left to top right across the sequence; **shading** – the circle alternates between black and white.

45 **c** **rotation** – the quarter circle rotates 90° anticlockwise as it moves around the box; **position** – the small circle moves in an anticlockwise direction from corner to corner around the box; **direction** – the arrow moves one and a half sides clockwise (so if it starts off in a corner, it will be next to a side in the following box); **shading** – the quarter circle alternates between white and black.
Distractor: **direction** – the arrow suggests the direction of movement of the quarter circle, but in fact they move in opposite directions.

46 **e** **position** – (a) the white square moves its own depth down the box, keeping in a central position between the left and right sides, (b) the black square moves up the box by its own depth, and also moves to the left by the same amount, (c) the

square with a cross moves up the box by its own depth and keeps to the left. In the missing square the white and black squares overlap; **shape** – the shape on the right alternates between a bullet and a triangle; **shading** – the bullet always has horizontal stripes and the triangle diagonal stripes.

47 **b** **position** – (a) the black triangle at the base moves one triangle to the left, (b) the white square alternates between the left and right sides of the box, (c) the vertical line moves gradually from left to right; **line style** – horizontal line at the top increases in depth between boxes.

48 **a** There are two alternating sequences, both of which follow the same rule. The missing circle is from sequence 2; **position** – (a) the circle is divided into four: sequence 1 with a vertical/horizontal cross and sequence 2 with a diagonal cross, (b) one shape in each sequence stays in the same place: in sequence 1 it is the quarter circle, in sequence 2 it is the rectangle, (c) the other three shapes move one-quarter clockwise every step of the sequence, missing out the quarter with the non-moving shape.
Distractors: **shading** – the shading visually makes the sequence more difficult to follow; **proportion** – the division of the circles into quarters also makes it more difficult to follow what is going on.

49 **c** **number** – (a) one small circle disappears on each step of the sequence, (b) one extra bulge on the cloud is added to every step of the sequence; **line style** – the style of the outer line changes in a repeat pattern of three: dashed, thick black, thin black; **shading/pattern** – the circles disappear in the order: all crosses, all black, then white remains (this can be worked out from looking at the last cloud in the sequence); **position** – the white circle is always in the bulge where the outline appears.
Distractor: **position** – the position of the outer line is random, it is only its relationship to the white circle that is important.

50 **d** **position** –the square moves one corner clockwise around the box in each step of the sequence; **rotation** – the triangle rotates 90° anticlockwise and moves one corner anticlockwise in each step of the sequence; **shading** – the shading follows a repeating pattern in every third box; **direction/style** – the arrow head changes in direction from up to down and then points in both directions in the third box, then the sequence repeats; **proportion** – the arrow head changes in size in every third box.

51 **d** **rotation** – (a) the diamond shape rotates 60° anticlockwise and moves to the next corner of the hexagon, (b) there is a pair of lines at a 120° angle that rotate independently with the white and black circles on the ends. This pair of lines rotates 120° anticlockwise in each step of the sequence, ending up overlapping the diamond in the empty hexagon; **position** – the extra two lines on this diamond alternate between left and right of the shape.

Distractor: **position** – the black and white circles do not swap places although it is difficult to tell without following the rules closely.

52 d **number** – (a) one line is added to the bottom right-hand side of the box as it moves across the sequence (counting all boxes) with top row adding a solid rule, (b) bottom row adding a bold solid rule, (c) the concentric boxes lose a bold outer line on each row; **shading** – the central box in the top row is black, in the bottom row it is white.

53 a **rotation** – the curved/zig-zag lines rotate 90° anticlockwise between the left and right-hand columns; **line style** – the line style of the horizontal and the curved/zig-zag lines swaps between columns; **shape** – the shape of the curved/zig-zag line changes between rows.

54 d **translation** – the small shapes in column 1 are translated into column 2 at varying sizes; **proportion** – (a) the top-left shape becomes the outer shape, (b) the top-right shape then second largest, (c) the bottom left the smallest shape so the bottom-right shape is in the centre; **line style** – the line style of the cross is used on the largest shape.

55 e **position** – (a) the right-angled triangle is always on the right-hand side of the triangle section, when the section is pointing downwards, (b) the black bar inside the triangle moves gradually towards the centre of the hexagon moving clockwise around the sequence, (c) the black bar outside the triangle moves towards the outer edge moving clockwise around the sequence. Distractor: **shading** – as both bands have the same shading it is difficult to distinguish the separate rules.

56 a **proportion** – (a) the proportion of the circle shaded alternates between one-third and full, moving around the outer edge of the grid, (b) the thirds will complete a fully shaded circle if they are all positioned together in one circle, (c) one sixth of the hexagon disappears in each hexagon moving clockwise around the sequence, **position** – (a) the zig-zag line moves clockwise around the hexagon as the sides are removed, (b) **shading** – the bands at the edges of each cell

change between rows, so the answer cell will match the cell on the bottom right.

57 e **shape** – the shape changes between columns; **translation** – rows one and two combine together to create the pattern in the third row: black segments that do not overlap remain black, black segments that do overlap revert to white.

58 c **number** – the number of diagonal lines in the square increases by one moving clockwise around the sequence from the left; **rotation** – the arrow rotates 90° clockwise and moves one corner clockwise between the square cells in the grid; **position** – the black and white circles both move one corner clockwise between the square cells in the grid. Distractors: **shading** – the movement of the circles is difficult to follow since there are extra circles that are not relevant in the triangular cells; **position** – the position of the arrows in the triangular cells also makes it difficult to follow the rotation in the square cells.

59 e **shape** – (a) the large shape changes between columns from a circle, to square to hexagon, (b) the small shape changes between rows from circle, to square to hexagon; **proportion** – the proportion of the shape shaded runs in a pattern from top right to bottom left; **position** – (a) the position of the small shape runs in a pattern from top left to bottom right, (b) the small shape is always diagonally opposite the quarter shading. Distractor: **shading/pattern** – the cross on the small shape is consistently placed on each row, i.e. in the top row it runs vertically and horizontally, in the second row between the corners and it is absent in the third row.

60 b **shape** – shapes match in a diagonal pattern top left to bottom right; **line style** – this works in a diagonal pattern from top right to bottom left; **position** – (a) the shapes in the top row are all at the top of the box, in the middle row they are in the centre and the bottom row all at the base of the box, (b) the circle moves from left to centre to right within its diagonal pattern, (c) the triangle moves from right to centre then left within its diagonal pattern, (d) the square stays vertically centrally within its diagonal pattern.

11+ Non-Verbal Reasoning Practice Papers 2 published by Galore Park